CHICKEN RAISING

THE BASICS OF COOP AND BREED SELECTION FOR BEGINNERS

PRISCILLA STEVENS

CONTENTS

The Beginner Gardener's Tool Checklist

A *must* read before you start your garden!

This Checklist Includes:

- 7 items you can't start a garden without.
- The highest quality items.
- Where you can buy those items for the lowest price.

Don't get overwhelmed when starting your first garden!

To receive your checklist, scan the QR code below!

INTRODUCTION

"There are blessings in being close to the soil, in raising your own food even if it is only a garden in your yard and a fruit tree or two. Those families will be fortunate who, in the last days, have an adequate supply of food because of their foresight and ability to produce their own."

— EZRA TAFT BENSON

When you decide to bring chickens into your world, you set a new tone for your life, a new rhythm to your day. You expand your role: You are now a keeper of poultry, a collector of eggs, a proud owner of independence.

Raising chickens in your own backyard can be a rewarding and often entertaining experience, as these fun creatures all have distinct personalities and behaviors that will bring joy to your day and a smile to your face. Add to that the benefits of having your own home-grown eggs and meat, and it is easy to see why backyard chickens are an appealing option.

It is not without its trials, and this is where doing your research ahead of time can pay off. Before you get your chickens, get your knowledge: Read this book to discover all the things that can catch you off guard in the early days of establishing your flock and all the ways you can prevent them.

A survey of experienced backyard chicken owners conducted by the blog on *The Happy Chicken Coop* found that the top three areas chicken owners wish they'd known more about were (Happy Chicken, 2019):

- Chicken Math
- Recognizing and Treating Common Illnesses
- Protecting Against Predators
- Preparing Your Table Fare

You may have heard of chicken math, probably uttered

by a chicken owner with a half-smile on their face and a knowing look in their eye. You might even believe this bizarre oft-joked-about phenomenon will not apply to you because you have self-control and a very clear mindset of what you want and how many chickens you need in order to achieve this. Famous last words.

As far as diseases and predators are concerned, chickens face unique challenges in both areas, and a sound foundation of knowledge will help you ensure your hens and roosters are healthy. Few things are worse for chicken owners than the feeling of helplessness that can come over you when one of your girls is sick and you lack the tools and knowledge to help them.

If the world of backyard chicken farming is riddled with so many pitfalls, is it worth the investment of time, effort, and money?

According to a 2013 article in *Mother Earth News,* eggs produced by backyard chickens contain, when compared to store-bought eggs (Caldwell, 2013):

- Half the cholesterol
- 25% more Vitamin E
- Approximately 20 times the omega-3 fatty acids
- 75% more beta carotene
- 67% more Vitamin A (Price, 2019)

- 25% less saturated fat (Price, 2019)

Factory Farm Chickens (2020) Retrieved from www.pixabay.com

The beta carotene gives the yolks a more vibrant coloring, the lack of cholesterol makes the eggs less slippery, and the healthier diet makes the eggs stronger.

In addition to the health benefits of eating more nutritious eggs, backyard chicken meat (if you choose to raise for meat, which is not necessary to keep backyard chickens) has been shown to contain far lower rates of hormones (Price, 2019). Factory farm chicken meat consumption has even been shown in a 2017 study to have a positive correlation to polycystic ovarian syndrome in women (Price, 2019).

Those are some inarguable health benefits to raising backyard chickens, but they are just the tip of the iceberg. As social creatures, chickens can be a delight to interact with and observe. They have a social hierarchy that is constantly evolving, they bond to humans, and they can interact well with children and family pets.

During the 2020 Coronavirus pandemic, the appeal of having backyard chickens has become apparent to a greater percentage of the population, especially in the United States, but the truth is that raising backyard chickens has been a growing trend for over a decade as more and more families seek out sustainable, independent living.

Now is a great time to take the plunge and begin raising poultry of your own. In the coming chapters, you will learn how to address every facet of starting your flock, with advice and tips from highly experienced chicken farmer Priscilla Stevens.

JacLou DL Three Chickens (2019) retrieved from www.pixabay.com

Stevens is a mother of two and a former manager at a highly successful Human Resources firm. In her early career days, Stevens never would have thought she would raise her own animals, let alone write books to welcome others into the joyful experience.

Her personal interest in homegrown food intensified when her younger child received diagnoses of numerous allergies and food sensitivities. This prompted her to investigate the truth behind the modern food processing machine: how meat, dairy, and eggs are raised conventionally. Uncovering many unethical practices inspired Stevens to raise her own animals in a natural and humane way.

Stevens has clocked more than five years of experience caring for her chickens, goats, and rabbits. She has learned from her mistakes as well as those of others, and she is ready to pass her hard-earned knowledge onto you, to ease your transition into chicken farming.

Whether you have already ordered chicks from your favorite hatchery or are still in the nascent stage of considering the risks and benefits of raising backyard chickens, this book will help you to feel prepared to make educated decisions, protect your flock, and herald your family into a new era of sustainable, ethical living.

1

THE ART OF TALKING YOURSELF INTO THIS

Something about the idea of backyard chickens appeals to you enough that you are actively considering the notion of starting your own flock. Whatever that is, whatever may have initially drawn you to this path, there are many more reasons that this will benefit you and your family.

TINY VERMIN

Wherever you live in the world, there are sure to be pesky, unwelcome insects that reduce your ability to enjoy your backyard. Many (but not all) of these can be mitigated by keeping free-range poultry in your yard (West, 2017).

Mosquitoes

Mosquitoes, according to a *ScienceDirect* article, are the deadliest animals in the world. They account for three-quarters of a million human fatalities per year (Ramsey, 2018) That does not factor in any animal or livestock deaths that occur. Mosquitoes transmit diseases such as malaria, dengue fever, Zika virus, West Nile virus, and chikungunya virus. That is barely a fraction of the myriad of mosquito-borne illnesses that can be transmitted to humans.

The good news is that chickens love to eat mosquitoes. This will reduce the number of mosquitoes able to reproduce in your yard, and in a short period of time you will see your unpleasant mosquito encounters drop precipitously.

This reduces the risk for disease, cuts back on the chances of anyone in your family experiencing an infection or allergic reaction to a mosquito bite, and makes it easier for you to enjoy your yard.

Ticks

Deer ticks and dog ticks are most notorious for carrying Lyme disease, but there are many other equally-pervasive illnesses that ticks can cause. Lyme itself can turn into a chronic, life-long medical condition in humans, impacting vital organs as well as emotional regulation.

The ticks that most often carry Lyme to humans, deer ticks, are often found in communities of mice, so it is important that you take measures to keep mice and other rodents out of your coop. We will address how to achieve this in a later chapter.

According to a 1991 study on chickens as tick predators, chickens allowed to free-range in a tick-infested area ingested an average of 81 ticks an hour (Hassan, 1991). That works out to 1300 ticks per chicken per day, assuming they spend 16 hours free-ranging!

Ticks not only carry Lyme (which causes a debilitating and life-altering set of symptoms in humans and can turn into a chronic infection) but several other diseases. Having chickens will help protect you and your family from all tick-borne illnesses.

Slugs

If you're looking for the sustainability afforded those who raise backyard chickens, odds are that you also maintain—or intend to start—a vegetable garden. Slugs can wreak havoc on a vegetable garden, leaving you with half-eaten greens and produce that hardly has time to grow before it is eaten.

The good news is that chickens love to eat slugs. They will decimate the slug population in your yard, giving your garden the freedom it needs to grow and yield food for your family.

Other Insects

Chickens feast on a variety of insects, from houseflies to horseflies and all sorts of other nuisances in your yard. Chickens will help keep them all at bay.

One matter that is worth noting is that chickens can become infested with fleas, lice, and mites. In a later chapter, you will learn how best to prevent and treat these infestations.

Lolame Chick (2019) retrieved from www.pixabay.com

LARGER VERMIN

As mentioned earlier, a chicken coop can attract rodents, especially if they can find a way into the chickens' feed. The rodents can, in turn, attract snakes. Smaller snakes may prioritize rodents and eggs for consumption, but the possibility of a larger snake going after one of the chickens does exist.

The good news is that when chickens see a mouse or a snake in their territory, their first instinct is to attack. So, while there are measures you can take to reduce rodent and snake access to your coop, be aware that many chickens are likely to kill any snakes or rodents that come into the coop (West, 2017).

CHICKENS AND CHILDREN

Now that we've talked about all the things chickens will happily eat, it is time to discuss children.

Chickens can and will eat them too, if the children harass them or if the chickens are not well socialized. Chickens that feel threatened, especially roosters defending their flock, will attack a child, dog, adult, horse...there are some highly entertaining video compilations online of roosters defending their flock.

This may seem like a reason not to get chickens if you

have little kids, but in truth this is a great learning opportunity for the whole family. Children can develop a lasting bond with their chicken friends, learn the responsibility of feeding and tending the flock every day, and develop new understanding and compassion for some of the world's creatures.

MESSY FOUL

Essentiell Two Laced Chickens (2020) retrieved from www.pixabay.com

Chickens produce massive amounts of waste per year. A single hen can make any from 30-45 lbs of litter per year (Harvill, 2019). Multiply this by the number of hens in your flock and you end up with quite a bit of fertilizer.

It has to be handled in the correctly: Too much chicken waste can burn the plants in a garden. Whether you choose to compost it for several months before adding it to spring soil, or whether you spread it in drops around the lawn to be sun-baked and then mowed over, chicken litter can do amazing things for your yard.

Why? First let's take a look at what's in chicken litter. Visible to the naked eye, it contains droppings, feathers, wood chips (or other coop bedding), and uneaten feed. What a mess!

On a nutrient level, however, this unique manure can work wonders for your garden and lawn. Some of the essential garden nutrients contained in chicken litter includes (Harvill, 2019):

- Magnesium
- Sulfur
- Calcium
- Zinc
- Copper
- Boron
- Iron

According to a 2012 article on *Holganix*, the six essential soil nutrients plants need are:

- Nitrogen
- Phosphorus
- Calcium
- Magnesium
- Sulfur
- Potassium

As you can see, chicken litter contains half those nutrients. It is a great no-cost way to provide healthy soil for your vegetable and flower gardens and help keep your yard both independent and sustainable (West, 2017).

PERSONAL GARDENING CREW

Chickens love to garden. Not only will they keep common garden pests, such as slugs and larvae, away from your plants, but they will also weed your garden for you. For this to work, you must have an established garden with larger plants capable of surviving on their own. Any seedling weeds that crop up will be quickly removed by the chickens. It will be eaten, turned into manure, and the nutrients will find their way back to your garden again.

In this way, farming with backyard chickens is becoming part of a great cycle of use, change, and re-use. There are several plants chickens should not be allowed to "help garden" with. These are addressed in Chapter 3 in the section on where to place your coop.

IT'S ALL IN THE FAMILY

These fascinating birds can serve another great purpose for your family: As companion animals. For people of any age who live alone, chickens can provide levels of emotional support that you might not expect. Often, well-socialized chickens will come running when they see you. Some will sit on your lap or actively seek out your company.

For the elderly, for those with anxiety, for exhausted parents in search of a moment of solace...chickens can and will provide comfort.

Some chicken keepers bond so deeply with their poultry friends that they invest in chicken diapers so that the chickens can free-range inside the house without making messes.

For younger family members, chicken farming is a great opportunity to learn responsibility as these birds rely on young hands to look after them day after day.

There is something fundamentally rewarding about the love the chicks repay your care with, and when a young child is first learning to prioritize the needs of the flock this reward will be bestowed upon them.

The best rewards are the natural result of doing good.

2

CHOOSING YOUR CHICK

When you select which birds to make part of your flock, the biggest distinction between them all is whether they are layers (bred for egg production), broilers (bred for meat-production), or dual-purpose birds (bred for both).

There are other elements to consider in addition to their growth rate and egg-laying capacity. Temperament, adult size, tendency toward broodiness, place in the pecking order, and vulnerability to illness are all factors to consider.

LAYERS

Layers tend to be heritage breeds, carefully intermingled over the centuries with various breeds with an eye toward early and prolific laying as well as barnyard hardiness.

Australorp

The record number of eggs laid in one year by an Australorp is 364 (Happy Chicken Coop, 2017), which makes it easy to see why this breed makes the list of best layer hens.

The breed has changed since Australian chicken breeders first experimented with a cross of various breeds in an effort to make a good egg-and-meat bird. Modern Australorps yield an average of 250 eggs per year, with a counterintuitive higher volume of eggs in winter. The majority of hens stop laying in winter, so this feature makes Australorps an appealing addition to the backyard chicken coop.

The earliest Australorps grew from a combination of Cook's Orpingtons and Rhode Island Reds in the 1920s. Experimenters added other breeds to the mix, looking for the hardiest bird they could make. Additional contributing breeds included Minorcas, White Leghorn, Langshan, and potentially Plymouth Rocks.

The result is a chicken breed with some clear positive traits and no real drawbacks in terms of hardiness or medical needs.

Breed characteristics include (Happy Chicken Coop, 2017):

- Prolific egg-layer
- The average age they start laying is five months
- Prone to obesity if not allowed to free-range
- Safe to free-range due to poor flying ability
- Cold-hardy
- Lay through winter
- Eggs are medium brown
- Life expectancy between 6-10 years
- Average weight around 6.5-8 lbs
- Need access to shade, due to coloring
- Hens tend broody
- Tend to be shy but will warm up socially over time
- Tolerant of most climates
- Friendly to people
- Roosters are docile toward people
- Get along well with other chickens

Surprisingly graceful for a larger-than-average breed, Australorps have a great deal to offer any backyard chicken farmer. In the United States, only the black variety is a recognized breed, but in Australia there are blue and white variations available. The breed standard Australorp has shiny black feathers with hints of green and purple in the sunlight. The comb and wattles are vibrant red.

Barnevelders

This is another friendly breed and is great for beginners and backyard chicken farmers. Developed in Holland in the latter half of the 19th century, this is another prolific layer. Though not quite as frequent a layer as Australorps—Barnevelders average three to four eggs a week (Happy Chicken Coop, 2018) while Australorps average five. They are another cold-hardy breed that lays through winter.

The breed, one of the heavier among chickens, features soft brown neck feathers and a double-laced dark brown and golden body feather pattern.

One fun trait Barnevelders are known for is their chattiness. While some other breeds that are conversationalists tend loud, Barnevelders have a more subtle voice and will happily greet you when you venture out to your coop or yard. They're known to be quite the story-tellers, with a pleasant enough voice to make it soothing and friendly.

Known for their shiny feathers, Barnevelders do well in any climate but are prone to overheating if the humidity is too high in summer. Shade trees provide a simple solution to keep your Barnevelder thriving.

A couple of notes on Barnevelders:

1. They are prone to Marek's Disease, so having them vaccinated against it as chicks may be wise.
2. They are *such* a sweet and docile breed that other chickens in your coop may be inclined to bully them.

With their inquisitive nature, great egg-laying rate, and friendly personalities, they are sure to be a great addition to any backyard flock.

Brahmas

meineresterampe Brahma (2019), retrieved from www.pixabay.com

Formerly one of the largest and best broiler breeds, Brahmas have become known as the gentle giants of the backyard chicken farm. They come in a variety of colorings, from light to dark to buff, and have a reputation of being loving even with the most enthusiastic of toddlers.

Historically, Brahmas played a key role in early North American farming. Known as the "king of chickens," the Brahma breed fueled a chicken-buying frenzy in the United States during the mid 19th century. The birds were famed for their size and egg-laying prowess and remained a popular breed until the advent of the Jersey Giant many years later.

They may not be the biggest broiler birds anymore, but they have plenty to offer the backyard chicken farmer. Weighing in at 8 lbs for the hen and 10 for the roosters (in the 1800s they weighed as much as 13 lbs and 18 lbs, respectively), it would be easy for them to dominate the rest of your flock, but they are too docile to worry as much about the pecking order as other breeds.

Brahmas average three to four eggs per week and actually lay *more heavily* in colder months, meaning that with a balance of these in your flock you'll have a steady egg supply year-round.

Because Brahmas have feathered legs and toes, they should be kept away from cold and wet places, as having wet feathers against their toes can make them prone to frostbite. Their feet can also develop mud balls, which can damage the toes if not kept at bay.

Their foot feathers can also snag on things in the coop or yard. When this happens, the wound can bleed an alarming amount. Coating the wound in cornstarch will staunch the bleeding and give the foot time to heal.

Isa Brown

Engineered by a French firm in 1978, the ISA Brown is a medium-sized bird with a prolific egg-laying rate and gentle personality that makes it a great family-friendly chicken (Happy Chicken Coop, 2018).

This breed was created to be great for egg-laying and they live up to the effort put into their genesis: They start laying as early as 16 weeks old and can lay up to 350 eggs per year. The only part of the year they don't lay is during their molt, which is brief.

Weighing in at a slight 5 lbs, the hens make for a great starter chicken because they are low-maintenance and companionable. They're ideal for urban and suburban residences where chicken farmers want to avoid noise complaints from neighbors, as they are a quieter, calmer breed.

Unfortunately, their intense laying schedule can take its toll on them. The chickens do better on a diet heavier in protein and calcium, to help them keep a steady supply of the building blocks they need to make their eggs. Oyster shell makes for a great calcium source (for them, and any laying hen) and should be given in higher availability following the ISA Brown's first molt.

Because the birds are a genetic hybrid, they do not have strong breeding capabilities. Their offspring tend to be weaker, less-prolific layers, and prone to kidney problems. First-generation ISAs can, themselves, have medical problems such as prolapse, reproductive cancers, and kidney problems. Most of these appear after the first two years of life, when egg-laying decreases significantly.

If you're looking for a loveable bird to welcome home, this breed can be a good addition to your backyard flock.

Leghorn

If you're looking for a breed with a colorful history, this bird has a fascinating past. Originally imported to the United States from Italy, the bird found popularity before it was *again* exported to the United Kingdom. Farmers there disliked how thin the Leghorn was and decided to cross-breed it with another chicken, making a bigger version of the Leghorn that was substantial enough to also be a broiler bird (Happy Chicken Coop, 2018).

That Leghorn was then exported to the United States again, leading to two different Leghorn lines in the US, with disputed and closely-guarded genetic standards. For those deeply invested in the modern chicken world, the Leghorn is part of an old rivalry stretching back to the 19th century.

Leghorns average 280-320 eggs per year, making them one of the best available backyard chickens when it comes to egg production. They tend to have a more calm, shy demeanor but will warm up to a social farmer in time. The hens range from 5-6 lbs, making them a medium bird, and the roosters average about 7.5 lbs.

If you decide to keep these birds, the only consideration is to keep Vaseline on hand if you live in a colder climate, as their combs are prone to frostbite. A coating of Vaseline will keep them safe from most cold temperatures and leave you with happy, healthy hens.

Marans

Marans are a fun breed with an interesting quirk: The

more they lay, the paler their eggs are. A Maran that only lays once or twice a week will yield a rich, chocolatey egg. A Maran that lays three or four times a week (or more) will lay more of a medium brown egg (Happy Chicken Coop, 2018).

The egg color also follows the cycle of the laying season: Marans lay from spring until their molt, with eggs becoming paler as the season wears on.

Anecdotally, the Marans is a popular breed among the British because it is supposedly James Bond's favorite egg. So, if you're on the lookout for the kind of egg that makes a super-spy strong and capable, the Marans is the way to go.

As a breed, Marans are not very prolific layers, averaging around three eggs a week. Males weigh about 8 lbs, with hens a little lighter at 6.5 lbs, making them one of the heavier birds in the layer category.

While their temperament tends docile, Marans are not cuddly birds and prefer to keep their distance from people. They also only rarely go broody, meaning that if you want more of your own Marans stock you should invest in and learn to use an incubator (covered later in this book). They are a bird that can thrive even in a smaller run, without the ability to free-range.

If you're looking for a hardy, content bird that will lay

steadily in the warm months and get along well with the rest of your flock, Marans can be a great investment.

Plymouth Rock

diapicard Two Chickens (2019) retrieved from www.pixabay.com

One of the oldest breeds in the United States, the Plymouth Rock was the standard for backyard chicken farming during World War II as the government encouraged families facing rations to farm their own chickens (Happy Chicken Coop, 2018).

As layers, they range around 200 eggs per year, laying more sparsely through fall and winter. They can be taught to go broody—perhaps by example from other breeds of hen—but are not prone to instinctually going broody. While egg productivity tends to go down around three years old, Plymouth Rocks have been known to lay for up to a decade.

Plymouth Rocks will be one of the most social hens in your flock, and will happily follow you around the yard. They are open, curious birds that love to socialize with humans and other chickens alike. They love to free-range, as it feeds their tummies and their inquisitive nature, but they can do well in an enclosed run so long as it is large enough.

Plymouth Rocks average 200 eggs per year, ranging from four to five eggs per week, and they lay through winter. They are great hardy birds whose only real health concern is that the roosters' combs and wattles need extra attention in the cold months. Careful application of Vaseline will prevent frostbite in these vulnerable body parts.

As an added benefit, Plymouth Rocks are one of the quieter breeds. If you live in an urban or suburban area where noise complaints might be a concern, they are a great addition to your flock as they will help keep the peace with your neighbors.

Rhode Island Reds

These have the claim of being one of the most popular chicken breeds in the world, both for industrial chicken farmers and backyard chicken farmers. Every Rhode Island Red chick you bring home to the safety of your yard is a chick that won't spend its life in the perils of industrial farming (Happy Chicken Coop, 2017).

The breed is so popular that there are even two statues of Rhode Island Reds in their namesake state.

They are hefty birds, capable of serving as broiler birds as well, with the males weighing in at an average 8.5 lbs and hens at 6.5. Their egg production is profound, as much as 300 eggs per year. They average five to six eggs per week. Over time, their size has gone down from full broiler size, in favor of egg production, and that vein of breeding has clearly paid off. Hens lay as early as 16 weeks and though they don't go broody often they make very protective mothers when they do.

Rhode Island Red roosters have developed a reputation for being aggressive, but keep in mind that there is variety within a breed. A well-socialized rooster can prove to be both protective of the flock and affectionate toward your family.

Medically speaking, they are among the most low-maintenance breeds, which is part of what makes them popular in industrial farming.

In the flock, Rhode Island Reds tend to be toward the middle of the pecking order. They love to forage and do well as free-range birds but can do well in a large enough enclosed run. They are *loud*, active birds, chatty both with people and with each other, and love to explore. While this makes them less than ideal if you're worried about complaints from neighbors, they make otherwise great pets and will absolutely ingratiate themselves into the hearts of your family.

Sussex

Sussex hens are another amazing set of layers, averaging up to five eggs a week and over three hundred per year. They only pause in their laying cycle for the very brief molting period and are unfazed by cold weather (Happy Chicken Coop, 2018).

This socially generous bird is known not only for its prowess in laying, but also for its friendliness in the coop and with humans. Roosters have an equally docile reputation, known for a mix of flock protection and human affection.

Sussex do well in an enclosed run if necessary, but are another breed that thrives with free-range of a larger area so that they can forage. They tolerate most climates well, but will need access to shade and cool water in the warm months. If you have any aggressive birds in an established flock, it may not be wise to add Sussex to your coop because they will likely fall to the bottom of the pecking order, so keep that in mind when considering when and how to include these in your chicken family.

The hens are prone to weight gain which will slow their laying. If you're interested in having heavier birds for the dining table, it is fine to let the Sussex gain weight, but if you want to maintain the rate of egg-laying it would be best to keep their body mass down. This can be done by limiting food or through free-ranging them so that they get more exercise and have to work harder for their food. Target weights are 9 lbs for roosters and 7 lbs for hens, so as you can see they tend toward the larger side of laying hen sizes.

If you're in the market for a great layer with the potential for a dual-purpose bird, or for a rooster even your kids can love, or for a hen that will happily climb onto your lap for cuddles, the Sussex may be the bird for you.

Wyandottes

minka2507 Columbian Wyandotte (2019) retrieved from www.pixabay.com

Wyandottes are a hardy breed of great layers who tend to be inquisitive birds and land towards the top of the pecking order. These birds are great brooders, so incubating may not be necessary—expect your flock to grow on its own if you get Wyandottes (Happy Chicken Coop).

Known for laying around four light brown eggs a week even through winter months, Wyandottes tend cold-hardy and healthy. Their rose comb (a differently-shaped, thicker comb) is safer than other birds' combs in cold weather. In the summer, they will need access to shade and cool water.

Their temperament in the flock tends to have them at the top of the pecking order. They're not particularly aggressive, but they know their minds and have no compunctions about putting other chickens in their place. With people, they are social and kid-friendly. They can do well in a large run but truly thrive when allowed to free-range. The roosters, because they are more opinionated, make great guardians of a free-range flock.

They are one of the larger-sized laying birds, with roosters and hens averaging 9 and 7 lbs, respectively.

With great protective roosters and hens that lay through winter, this breed is certainly worth considering for your backyard flock!

BROILERS

Broilers tend to be more modern than the heritage breeds, as industry farmers work to make more and more fast-growing, meat-heavy, delicious meat.

Bresse

The Bresse (pronounced *bress*), is a meat bird with a curious history. Reputed to be the most delicious of all the meat birds (it was bred to metabolize nutrients a certain way to improve flavor, and bred for small bones and heavy meat to improve meat sales per chicken (Happy Chicken Coop, 2019).

True Bresse are only available in France, already slaughtered, and edible for a hefty price. Anecdote exclaims that the experience is worth the cost.

In the United States, you can purchase living chicks from US-bred-and-raised Bresse chickens descended from the famed breed of France. Ostensibly there isn't much difference between them, but don't tell that to the breeders in Bresse, France, or to the gourmet chefs in Paris who sell the meat.

A Bresse chicken will be white, with blue legs and a single comb. There are techniques the French follow when deciding at what age to cull the chickens, and the diets are arranged around the culling. For the backyard farmer, you can expect your chicken to reach a weight of about 6.5 lbs if you allow it to grow to adult size.

You can cull as early as four months and still get a substantial amount of tender meat.

Cornish Cross

Cornish Cross chickens, another mostly-all-white breed of broiler bird, have the auspicious claim of being able to grow to 6 lbs in just six weeks. This means if you're looking to grow a lot of freezer meat in a short period of time (such as a pandemic), these might be the broiler birds for you (Happy Chicken Coop, 2019).

They require a special diet since they grow so fast: You will want to keep them eating plenty of protein and fats to ensure their steady growth, Additionally, you will want to offer food in strict 12-hour shifts to give them a balance of time to grow and time to eat.

Because Cornish Cross grow so quickly and yield such delicious meat, they are one of the most common birds found on supermarket shelves. Adult hens weigh in around 8-9 lbs and roos tend between 9-11 lbs, making this a great breed for table fare.

Freedom Ranger

These birds grow more slowly than breeds like the Cornish Cross, so it's not quite the same frenzy of chicks-to-chickens-to-table that other breeds can be, but they're still mature enough to harvest far faster than heritage chicken breeds (Happy Chicken Coop, 2019).

They reach their harvest weight within three months, topping out around 6 lbs, and they yield succulent and tender meat that is lower in fat and cholesterol than faster-growing broiler breeds. Expect juicy and delicious table fare from this bird.

Freedom rangers tend to be red or tri-color, with full combs.

Jersey Giant

Developed in the late 19th century in New Jersey for the sole purpose of being broiler birds, Jersey Giants make a good moderate-growing table fare. You can find them in black, white, or a slate gray color, although the lighter birds tend to weigh less (Happy Chicken Coop, 2017).

An adult male Jersey Giant will weigh in around 12 lbs at maturity, with females slighter at around 6.5 lbs.

The advantages of raising Jersey Giants are numerous: As moderate growers, they tend to be lower in fats and cholesterols than faster-growing birds. The hens tend to be prolific layers, bringing in around two to four eggs a week (150-200 per year), and they are one of the breeds that lay through winter.

You could conceivably have a backyard flock of *only* Jersey Giants and manage to keep a steady supply of roosters for meat, broody hens making more chicks, eggs for consumption, and one or two mature roosters to watch over the flock and sire the young.

In that way, the Jersey Giant may be among the most ideal dual-purpose chicken breeds you can invest in.

Orpingtons

Orpingtons are showy birds—literally. They tend so docile and so impressive to look at, that egg production has fallen off in recent years in favor of breeding for appearance (Happy Chicken Coop, 2017).

Despite this, the breed continues to be decent layers. Males and females both weigh in around seven to eight pounds, and they make for fleshy and delicious table fare.

3

NO PLACE LIKE HOME

When you raise backyard chickens, one of the most important facets of their quality of life (and yours) is their living arrangement. Whether you have half an acre or a hundred, where your chickens live can influence their health, happiness, and productivity.

Some people choose to go all-out, designing and painting miniature cottages with welcome signs and stenciled flowers, or large walk-in coops with plenty of room for chickens to stay warm in the winter. Others take a more minimalistic approach, planning on chickens that free-range year-round and only need a small, secure space for their chickens to sleep.

Whichever category you fall into, or anywhere on a

spectrum between them, some basic coop essentials that will help you and your chickens have happy, comfortable lives together.

TYPES OF CHICKEN COOPS

Just like homes for humans, which can range from flats to apartments to houses to mansions, chicken coops come in a variety of shapes and sizes. Each one has a purpose, as well as benefits and detriments.

Depending on your needs and budget, any one of the following coop types could set you up very nicely for a long-term commitment with your new flock (Stratton, 2013):

Elsemargriet Chickens Roosting (2020) retrieved from www.pixabay.com

Stationary Coops

These are the more conventional, traditional coops that you might imagine from movies or books: A building where chickens live. It doesn't move. The floor may even be finished, the walls secure.

There are numerous benefits to having a stationary chicken coop. For one, the chickens know where home is. This means they can come and go and will only free-range within a certain distance of their coops. While chickens are not territorial in the same sense that other animals can be, knowing their 'range' can help them feel more relaxed in your yard.

Another benefit to stationary coops is that you do not need to continually re-secure the coop against predators. Once you've established a technique that works to preserve safety for your flock while allowing them space and ventilation, you don't ever have to change it again. This saves time and effort.

Because your coop never changes, your routine never changes. Any cleaning supplies you need can be kept right there at the coop and will be readily available when you're ready to clean. You can keep composting bins, scrap barrels, rain barrels, and other useful accessories right nearby with ease.

Not moving the coop also means it doesn't face wear and tear. While mobile coops are designed to be moved, it still places a strain on the joints that stationary coops simply never face.

Finally, you can easily run electricity and water to a stationary coop through a secure underground conduit. This is impossible to do with mobile coops, where you would rely more on extension cords.

For a finished, secure, easy coop, the stationary option may be the best course of action for your family of birds.

Semi-Permanent Coops

These are similar to permanent coops although they tend to be a little on the smaller side, comparatively. Unlike a permanent coop, which may have a paved floor or be fixed to the ground, these coops *can* be permanent or can be relocated as needed around the yard. They tend to be smaller than permanent coops for better mobility.

As with the permanent coop, there are a handful of advantages to using this system. One is that you only have to re-secure against predators if you move the coop. Relocating the coop seasonally, or once a year to move between summer and winter locations, limits the work you have to do.

Keeping the chickens in one place most of the time will help them to remember where home is. Some chickens will even put themselves to bed in their coop at night!

The smaller, semi-permanent coop has the added advantage that it is designed to be both easy to clean and easy to harvest eggs from. This simplifies your life: Just pop out to the coop, grab the eggs, wipe everything down, and you're on your way.

Similar to the permanent coop, it is possible to run permanent electrical and water lines out to your coop locations, if desired, without having unsightly electrical cords above the ground. A trench and some steel conduit can make this process easy and keep your coop warm in winter months.

This sort of coop also has one disadvantage of note: If you leave it, say in mud or soil that attaches to it, and then try to move the coop you may damage the structure just breaking it free of the ground. Care should be taken with the frame and base to ensure it is ready to move before dislodging it.

Raised Coop

A raised coop can come in permanent-style, semi-permanent, or tractor-style, but regardless of that there is one trait unique to the raised coop: The actual 'house' portion of the coop sits off the ground. There are several advantages to this design.

As mentioned above, structures of permanent and semi-permanent coops can fuse to the ground in places where there is mud or even just clay-like soil. If the physical structure of the coop is against the mud, this can lead to rotting of the floor, beams, and joints over time, leaving the mesh vulnerable to attack from predators. Lifting the coop of the ground protects your structure and your birds.

With your coop off the ground, digging and burrowing animals (predator or foragers) will not be able to chew their way into the coop. This protects your flock from both injury and disease, as well as helps keep the feed clean.

If you cannot free-range, having the coop structure itself off the ground means you have more square footage at the ground level of your run. This gives your birds more space to move around and keeps the run cleaner. Under the coop is a great place to set up a dust bath for the chickens, as it will be protected somewhat from the elements *and* will be shaded.

The higher coop will make it easier to access eggs and clean out the coop, just by not having to bend over. While this may not be a consideration now, remember that most breeds live six to ten years or even longer, and your body's needs may be very different in a decade. It is good to plan ahead.

To prevent digging predators from clawing their way into the run, it may be wise to line the bottom of the run area with a wire mesh or hardware cloth to help protect your flock.

Tractor Coop

This is a (typically smaller, though not always) coop that has handles on one end and wheels on the other, like a garden cart you can pull around. This can be useful if you cannot free-range your flock but want it to have a full range of the yard, or if you need to move the coop around seasonally or for other reasons. It comes with some good perks.

The ground around a permanent coop, whether your chickens free-range or use a run, is likely to get muddy without proper care. Even *with* proper care, it can take a lot of work to keep mud from becoming a problem for your chickens. With a tractor coop, however, all you need to do when the ground gets bare is to move the coop somewhere else. Problem solved!

Moving the chickens around every day will not only give chickens access to fresh ground (and whatever fresh grubs they can dig up) daily, it will also keep your lawn in constant fresh fertilizer.

A final boon of this coop design is that when the location of the coop changes, it is often difficult for predators to establish any kind of habit when it comes to your coop. They cannot dig a hole for a week before they find a way in, or investigate the same corner every night. Your coop is more of a mystery, which helps keep your birds safe.

INGREDIENTS FOR A GOOD COOP

Whether you will ultimately free-range your flock or keep them in a run (or any mix of both), there is no doubt that the coop itself has some essential facets that must be included. You can build it, you can buy it online, you can even order coop design plans from some do-it-yourself and craft product websites. At the end of the day some fundamental elements that must be universal to your coop (Timber Creek Farm, 2016).

You will need a sturdy structure that offers protection both from weather and predators. If it is permanent, it should be placed somewhere it will not rot over time—away from flood areas and poor-draining soil. Adding a mesh lining along the bottom of the run will help prevent predator access.

The coop should be well-ventilated. Screens and windows help maintain overall flock health and reduce odors. Chickens without access to a clean well-ventilated coop can develop respiratory issues, and a closed-off coop is a breeding ground for parasites and rot. Any windows or mesh screens incorporated into your coop should have a way to be closed off against the coldest nights of winter and the windiest nights of summer.

Unless you are strictly raising broiler birds that are all cockerels, your coop will need nesting boxes. These are small semi-enclosed areas that provide shelter and bedding for the hens while they lay. These should be placed in the shadiest area of the coop or run and be accessible to hens throughout the day and night.

Anecdotally, even if you have a nesting box available for each of your hens, they will likely all use the same box and fight over who gets to use it first. It can be a little like having one bathroom and three teenage daughters who have to get ready for school.

All chickens, regardless of gender, will need a perch or roost bar. This should be ten linear inches per chicken in your flock. Many people mistakenly think the roost should be round, but this is inaccurate and can even hurt your birds. They need a flat roost that enables them to tuck their toes safely under their feathers in colder months to prevent frostbite.

Somewhere in your coop or run (one in each place is ideal) your chickens will benefit from having a dust bath. This bath, made of dry soil or sand with either ash or diatomaceous earth, helps prevent mites and parasites and the chickens love to preen and clean themselves in the dust. It is truly a pleasure to watch and it brings them much joy and improved quality of life.

Every coop needs a feeder and a waterer. There are numerous ways to do this, from trough feeders with screened grates to simply having a broad-based dish out where chickens can access their feed. Because of the risk for parasites such as worms and coccidiosis, it is not recommended that you sprinkle feed on the floor of the coop or run for your chickens. Doing so can also attract unwanted vermin such as mice and rats to your coop.

The waterer should be off ground level to help prevent contamination and it will need to be cleaned and replaced daily. Keep that consideration in mind when deciding on the type and placement of your waterer.

If you live anywhere that sees extremes in temperatures, you will want to look into insulating your coop against heat and/or cold. Options if you cannot insulate include running electricity to the coop so that you can have a heater or cooler for the chickens. Regardless of which method you use, keeping your flock at a comfortable temperature is critical for long-term chicken health.

Remember also that a coop that is too big will be difficult to keep warm in colder months, but a coop that is too small will be difficult to keep cool in warmer months and will encourage the spread of disease in your flock. A good rule of thumb is three square feet per bird inside the coop and 25 square feet per bird inside the run. Also remember that smaller coops must be cleaned more often to maintain flock health. A coop that doesn't have a minimum of three feet per bird will likely need to be cleaned daily to keep down odors and risk for disease.

If you include all these elements, whether you purchase all or part of them or build them from scratch, you will be well on your way to having a healthy, happy chicken flock.

TO BUILD OR TO BUY

When planning the place your flock will call home, you essentially have three possible starting places:

You can buy a prefabricated coop in a box, which is intended for you to assemble with minimal effort and time invested. These typically include pre-cut, pre-drilled pieces ready for you to build into a sturdy coop.

Another possibility is that you can purchase (or already own) a shed that you convert from a tool storage area to a chicken residence. This kind of coop can only be a permanent coop—you won't be relocating a shed around your yard seasonally.

The final option is to build something yourself, from scratch. The internet is full of websites where you can buy or borrow coop design plans to help you know what kind of coop you want to build.

Before you make any big decisions about the financial and time investment involved in whichever avenue you pursue, now is a good opportunity to read up on the pros and cons of each method.

Prefabricated or Pre-Fabulous?

The clearest advantage right out of the gate with the prefabricated coops is that it requires minimal planning and effort from you. You get the box (hopefully with no pieces broken or missing, which can be an unfortunate but rare outcome of going this route) and you assemble the coop from the instructions provided.

You can probably also trust these coops, right? A company that makes chicken coops knows chickens better than you do, especially if you're a first-time backyard farmer. Unfortunately, this isn't necessarily true.

Some of the plastic coops are great for security but offer no ventilation. Plastic can get brittle in cold weather and break.

Wood coops can be of flimsy construction, with loosely-fitted mesh or hardware cloth ventilation that is easy for predators to tear out of the wood frame. The wood itself may be of a material that is not stalwart enough to stand up to weather where you live.

All of that said, there are some very well-made coops available for purchase online and at supply and hardware stores. It is key with pre-fabricated coops that you read all the reviews, especially those about the sturdiness and hardiness of the materials. Look for any written a year or two after purchase to gauge how well the coop has held up over time

The more variety in materials, the more likely the coop is to hold up to various types of weather. Many prefabricated coops include metal roofing, wood framing and housing, and mesh or hardware cloth screening.

Look for coops made of stronger woods. You don't

want to invest in something pricey only to discover it is made of painted particle board. Tropical hardwoods and redwoods are rot-resistant and tend to make the best quality wood for coops.

Remember as you shop that ventilation is critical with prefabricated coops. Many plastic coops offer no ventilation whatsoever and are not a wise purchase.

Modifying a Shed

Sheds can be quite roomy and comfortable for chicken coops, and come with the added advantage that you can just walk straight in an adult-height door to do your cleaning and tend your flock. Add to this, that you can get out of the weather while caring for your hens and roosters, and it's easy to see the appeal of modifying an existing shed.

Typical prefabricated sheds and outbuildings on your property come in one of three materials: wood, plastic, or metal. Each has advantages and disadvantages:

- **Metal**: Metal sheds have the advantage of being cheaper than wood and plastic, and quick to assemble. It is more difficult to install roost bars, ventilation, and insulation. Metal sheds also cannot stand up well to shear wind so if you live in an area prone to high winds it would be wise to forego metal sheds or set up a wind barrier. Many metal sheds also come with a framework for the floor, making it easier to install flooring.

- **Plastic**: Plastic sheds tend to be a little more pricey than metal, but they come with increased stability and wind-resistance. It is easier to drill through the walls to install roost bars and ventilation windows, if the shed doesn't come with windows (many do). They are not as easy as wood to insulate but won't get too frigid or too overheated nearly as quickly as a metal shed would.
- **Wood**: Wood sheds are the easiest to convert and insulate. The design of the wall framework is conducive to including insulation that will keep the coop cool enough and warm enough depending on the month. It is easy to drill into walls to add shelving, nesting boxes, and flat roosting bars.

Regardless of which material shed you choose to modify into a nice big chicken coop, one of the best facets of using a shed is the ease with which you can provide a finished flooring of some sort. That will help keep the chickens warm in winter and greatly facilitate cleaning.

Built from Scratch

Building your own coop gives you the most flexibility to customize and make a coop that fits your specific needs. This is a good time to familiarize yourself with the important factors to incorporate into your coop and the building process.

Make sure the wood you use is not toxic or quick to rot. Most notably, avoid using cedarwood in your coop, either as wood or as chips for bedding. While cedar is good for reducing the spread of mites, parasites, and disease, it can cause respiratory troubles in your chickens and make them more vulnerable to health issues.

You will also want to avoid the use of corrosive metals in your coop. Not only will they compromise the integrity of the structure, but they can also oxidize and distribute unwanted rust and flakes of metal around the coop. Since chickens will eat flakes of anything and you don't want them consuming metals, it is wise not to have bits of corroded metal sitting around the coop.

Remember that your coop should allow three square feet per bird. This gives them ample space without risking exposure to cold temperatures. This means that when you plan your coop, you'll need either to purchase a flock to fit the coop or you need to build a coop to fit the flock you've already ordered.

For example, say you build an 8 ft by 10 ft shed. That leaves 80 square feet of space, which is enough to accommodate approximately 25 chickens. Now let's say you've only ordered ten chickens. What do you do? You may eventually decide to grow your flock, so you want to keep that nice spacious shed and block off part of it to use later. Keep the actual portion of used shed space down to about 30 square feet for your flock of ten.

Once you use the guidelines above to determine how big you want your coop to be, it's time to find plans! Plenty of resources exist online, populated with chicken coop design plans. You'll want to take time to study a variety of these and see which elements you wish to incorporate into your coop, and how you want to budget things. There is no "one size fits all" approach and you can build you coop as simple or as fancy as you like. Some of the plans are available for a small fee, but you can find others for free without much effort. Social media also has backyard chicken farming groups that can offer tips and examples of their own coop designs.

Have fun! Planning and building your coop should be an exciting experience. If you're stressing it, take a step back and remember this doesn't have to be perfect: just functional.

WHERE TO PLANT YOUR FLOCK

Pexels Rooster on Table (2020) retrieved from www.pixabay.com

Consideration must also be given to where you want your chickens to live. Some of this is based on their need and some is based on yours.

If you live in an urban or suburban area, you will want to check zoning laws. Some areas will not allow roosters, some limit the number of chickens, and some even dictate where on your property a coop can be constructed. Other factors for urban/suburban regions to consider are covered here.

You'll want to factor in spacing and fencing between yards in your area. If there is no fencing between your yard and your neighbors' yards, or if the fencing is low enough for a chicken to fly over, you will want to keep the coop as far from the edges of your property as possible.

Another consideration is that different fencing materials can change the environment around your coop. For example, mites aren't likely to grow between a chain-link fence and a coop, but they may find a comfortable place to hide between a wooden coop and a wooden fence. If your line fence is wood, that will also be more difficult to clean than vinyl or metal fencing.

You'll want to be aware of what plants your neighbors grow near their property lines. Many plants are toxic for chickens and should be avoided. These include (but are not limited to) daffodils, foxglove, morning glory, yew, jimson weed, tulips, lily of the valley, azaleas, rhododendron, mountain laurel, monkshood, amaryllis, castor bean, trumpet vine, nightshade, nicotiana, and tansy. If any of your neighbors grow these along the fenced area, it may be wise to keep the coop far from that neighbor's property line.

The size, shape, and features of your yard are especially important if you intend to free-range your flock. You'll want to keep the coop away from the road, as far as possible without causing problems with neighbors. If you have any bodies of water such as ponds or swimming pools, you should keep the coop distant from those because chickens are terrible swimmers. If you have any other animals in your yard, especially dogs or a flock of a different type of fowl, you'll want to keep them separate in your yard. Any animals that might be aggressive toward your flock or spread diseases to your birds should be isolated physically.

You'll want to factor in drainage and shade as well: The coop should be set in soil that drains well, to prevent rot and disease spread, and you'll want some portion of the coop and run to be in the shade to give the chickens somewhere cool to rest during warm summer months. Nesting boxes for your birds should be located in the shade and secluded if possible.

When selecting a place, don't forget to consider water and electrical access. Having the coop in easy range of the water hose will make maintenance, cleaning, and watering easier. Similarly, having access to electrical power means you can heat and cool the coop as needed seasonally as well as heat the water dish so it does not freeze in winter months.

Similar to placing your coop for access to the conveniences of modern technology, another facet to weigh is the ease of access for the coop. Many people start their flock in the spring or summer months, when going outside to check on the coop and birds is a pleasant and comfortable experience. Keep in mind that in the winter, it may not be so enjoyable. You'll want your coop to be somewhere you can do your chores quickly without too much exposure to extreme cold in winter.

If you aren't sure where you should keep your coop right away, one solution might be the tractor-style coop discussed earlier in this chapter. You can move it around your yard until you have a sense of where the best coop placement will be. Then you can invest in constructing a permanent coop in that spot, satisfied that you will not regret the coop's location.

A SUMMARY OF COMMON ROOKIE ERRORS

There are a handful of troubles that are easy to avoid right away if you take the time to be careful.

The more you can do to protect against predators early on, the better off you and your flock will be. Remember to keep your coop off the ground, if possible, to prevent predators from digging and chewing their way in.

There's an easy-to-imagine scenario, a mix of bad luck and carelessness, that can befall your flock. Imagine you have a few days where you don't inspect your coop. This gives time for a predator to find a way to tunnel under the coop. From there, at night, they can chew their way through the base of the coop and attack your vulnerable chickens while they sleep. This sort of thing is why it is so important to be vigilant and aware of your coop from day to day.

Other ways to deter predators include adding a layer of hardware cloth over screens and across the base of the run. This will help prevent predators from chewing or clawing their way in.

Few things will protect your flock more thoroughly than a good latch on your coop. Remember that clever predators such as raccoons can open a simple latch, so a lock might be advisable depending where you are located.

As mentioned earlier, the size of your coop is also important: Too big, and your chickens risk frostbite. Too small, and they risk disease. In the same frostbite-avoiding vein, do not forget that roost bars should be flat.

Don't forget weather: Windy areas need sturdy buildings; hot or cold areas need insulation, heaters, coolers, and access to water heaters in winter; hot areas need access to shade and should not have a metal structure as they are prone to retaining heat in the sunlight.

Finally, make sure your coop is easy to clean and accessible year-round for cleaning, flock inspections, and egg collection.

Following those steps will ensure that you love your coop for years to come as you integrate your birds into your yard and lifestyle.

SAMPLE BUILDING PROCESS

Part One: Planning

In the early stage of the coop design, you'll want to toy with ideas for a few days to be sure you're investing your time and money toward the right efforts.

You'll want to calculate the square and linear footage you will need to accommodate your flock. This is ten linear inches of roost bar per bird, three square feet per chicken inside the coop, and 25 square feet per chicken in the run.

With the coop size, you want to keep it pretty close to the three square feet per bird, but the run can certainly be bigger than 25 square feet per bird. Bigger runs make for happier birds.

Once you know the footprint of your coop, decide where it will go and how permanent it will be. Tips for choosing a location were covered earlier in this chapter, so be sure to refer back to those as needed while selecting your placement.

Clear the area and prepare it for building your coop. Once that is complete, gather your materials and supplies so that you are ready to build. This should include the screws, nails, and wood that are specific to the coop design you will be using. You will also need a screwdriver, saw, sandpaper, level, hammer, measuring tape, pencil, right angle, primer, paint, and paintbrush. An advantage of gathering your supplies ahead of time is so that you do not have to interrupt your work mid-build in order to run out and purchase something on the fly.

With your materials assembled, you are ready to begin the exciting process of building your very own chicken coop.

Part Two: Building

1. Start by assembling the walls that will run the length of the coop. The framework for these should be level with the right angles and should follow the instructions provided by whatever plans you are using.

2. Assemble the end walls/depth of your coop. Make sure all pieces are level and right before attaching them to the side walls.
3. If you intend to have a framework under the floor, add that at this time.
4. Put together the roof trusses. You should angle these to your preference, depending on how steep you want your roof to be. Remember that in areas that see heavy snow, steeper angles are better for the roof.
5. Once the frame is assembled, gather your wall, floor, and door pieces. Cut them to size and make any necessary cuts for window openings.
6. Attach the walls and floor into place.
7. Put roof pieces in place.
8. Install your door and window pieces
9. Sand off any lingering rough areas and then coat your coop in a layer of outdoor-friendly primer.
10. Paint your coop.
11. Install any roofing slates or tiles you wish to use.
12. Enjoy the coop!

4

HELPFUL HINTS AND TRICKS

GREAT DEBATES

In the backyard chicken farming world, there are some areas of controversy that are more matters of opinion than cases of a right way and a wrong way. Among these are:

- Whether to free-range your chickens or keep them in a run
- Whether to vaccinate chicks (and if so, which things to vaccinate for, and when)
- The risks and benefits of roosters in your flock
- Using medicated versus unmedicated feed
- Culling versus giving away spare roosters

Rather than get into the details of the debate or make a case for either side, this section of the chapter will list the pros and cons to help you make an educated decision about your flock, your backyard, and your family's needs.

Home on the Range

Free-ranging your flock, in simplest terms, is letting them go where they want (within the limitations of your property) during daylight hours (Elise, 2019).

There are some strong arguments in favor of free-ranging your flock. For one, your flock will be able to forage bugs (including ticks and mosquitoes) from all over your yard, not just the area contained within the run. Chickens can eat hundreds of ticks and mosquitoes an hour!

Your chickens will get better exercise if you free-range them than if they are (literally) cooped up in a run all day. Because they'll be in fresh, clean areas of the yard throughout different parts of the day, they are also less likely to be in muddy or dirty conditions prone to spreading disease. This is not to say that chickens kept in a run are more prone to disease: More that if you keep them in an enclosure you must do more work to get the same desired clean effect.

Chickens that free-range tend to get along better as

well. With more room to spread out, this should be no surprise that they have less to fight over. If they don't get along, they can simply spread a little further out until tempers calm. Though they still have a pecking order, it can be less violent when the hens are free to get further away from one another.

It is also easier for your flock to self-regulate: They go to areas with food when they are hungry, with shade when they are hot, with dust baths when they want to clean themselves. They can sunbathe or seek out shade, and they can choose to be social with you or with each other, or seclude themselves under a hedge. In short, your chickens have more freedom of choice, and since they know what they want better than you do it is a logical conclusion that chickens get more of what they want when they free-range.

That being said, there are some definite cons to having chickens roaming around your yard.

The clearest of these is the risk from predators, which can include birds of prey, foxes, raccoons, skunks, possums, and even neighborhood dogs and cats. While you can take extensive effort to predator-proof your yard and add flashing to your trees to deter birds, the truth is that it is much easier to secure a run than a whole backyard.

When you bought your house, you might have gotten a handy drawing of your property, which shows the borders, maybe the location of sewage and electrical lines, where your water gauge is, and exactly where your property ends and your neighbors' property starts. The road is clearly labeled on this drawing as well.

wilma polinder Chickens on Fence (2020) retrieved from www.pixabay.com

Your chickens did not get this drawing. Fences and hedges—even roads, to some chickens—are challenges. If your neighbor has a sandbox and your chickens want to be there, they *will* find a way.

You may need to trim your chickens' wings to keep them in the boundaries of your property, especially if you have an ornery neighbor who doesn't enjoy the idyllic clucking in the morning. If you free-range, you will need a way to keep your flock in your own yard and out of the road. You will also need to accept that maintaining a pretty flower garden or a refuse-free porch and patio area will be nearly impossible with chickens on the loose.

The unfortunate drawback to trimming wings is that it makes it more difficult for birds to escape predators. What you end up doing is exposing them to added predator risks by letting them free-range, and then you make them more vulnerable by trimming their wings so they can't escape.

In the end, you are the only one who can decide the risk-benefit outcome for your flock. If you free-range, there is a level of associated risk, but the benefits of fresh grubs all day are impossible to deny.

Vaccinations

As passionate as parents can be about the morality and safety of vaccinating their human children, this amazing debate carries over into chicken farming as well. Everyone has an opinion. Studies are rampant online, as are conspiracy theories.

The truth is that we have the great fortune of being able to decide this one for ourselves. You can choose to vaccinate, or not, and no one can call you out on this decision.

However, you should have all the information you needed in order to make a well-thought-out, educated decision about what to do with your flock. You may choose to utilize all available vaccinations, choose some like selections at a buffet based on your needs and where you live, or eschew vaccinations altogether.

This is a brief overview of available vaccinations, and the fundamentals of when, how, and why the vaccines should be given. Vaccinations are available for many poultry diseases that can negatively impact your flock, including (Cackle Hatchery, 2018):

- Marek's Disease
- Has to be given on the first day of life, or not at all
- Chicks vaccinated against Marek's *can* be shipped through the mail
- Newcastle Disease
- Chicks vaccinated against Newcastle cannot be shipped through the mail
- Requires boosters until the 18-20 week age vaccine which is more potent
- Infectious Bronchitis
- Often paired with Newcastle Vaccine
- Infectious Laryngotracheitis
- Can be vaccinated against *during* an outbreak but requires government oversight

- Fowl Pox
- Does not protect against all types
- Not needed unless a farm is prone to fowl pox
- Fowl Cholera
- Should not be vaccinated against unless it is a problem where you live

The general debate, when it comes to vaccinations, is what purpose they serve: If you have an isolated, small flock of birds that will hatch its own chicks and be self-sustaining, what disease risk is there?

The argument isn't misplaced or invalid, so long as you take care to quarantine any new birds or chicks for 30 days and so long as you can ensure that other chickens will not have access to your birds.

Chapter 7 will discuss each of those diseases, as well as a handful of others, in more detail. Some backyard chicken farmers choose to vaccinate against specific diseases while ignoring others, some choose to vaccinate against everything, and some choose to play the odds.

It is worth noting that if you wish to sell certified organic eggs, you cannot vaccinate.

To Roo or not to Roo

klimkin Chicken Flock (2020) retrieved from www.pixabay.com

One of the biggest areas of debate among backyard farmers is whether or not a rooster will benefit your flock. There are so many elements that factor into this decision that it is worth further study here (Green Willow Homestead, 2017).

The quickest and simplest deciding factor is whether roosters are allowed by zoning laws in your area. If they are not, and you attempt to have one, you risk losing your entire flock. Since roosters are known for loudly declaring their presence several times a day, efforts to hide them are ill-advised.

If zoning laws are not a consideration, there are other elements to consider. One perk to having a rooster is that they are highly protective of their flock, which will make it easier to safely free-range your birds. One detriment to having a rooster is that they are highly protective of their flock, which will make it difficult for you to tend the hens if the roosters decide you are any kind of threat to them. Two sides of the same coin: If you free-range, the more aggressive a rooster you can find, the better; if you want a relationship with your flock, the calmer breeds of roosters are your best bet.

Spacing issues are an important factor when considering whether to have a rooster, not because roosters are bigger than the hens in your flock (though they are) but because a rooster in the flock, plus a broody hen, means a bigger flock in a matter of weeks. This is chicken math at its finest—say you have one broody hen sitting on twelve eggs and suddenly find yourself with eight extra chicks. Where will you put them? What will you do with the 50% of the chicks that are roosters?

You absolutely must be willing and ready to cull or sell the extra roosters, or having a fertile flock is not right for you. If you are ready for that facet of rooster-ownership, it is a fabulous way to organically grow your flock for only the cost of chick starter feed. It also means you can grow your flock without risking the introduction of new diseases to your birds.

If you choose to keep a rooster, you should spend as much time as possible socializing with him when he is a cockerel. Teach him to trust you, to see you as a protector. Have him eat from your hand. Greet him first when you visit your flock, so he knows he is the special one. The pride that instills will help ensure he values the work he does to protect his flock, because he knows you value it too.

Don't stop socializing with him once he matures, either: Continue greeting him daily, feeding him from your hand, and praise him. If he ever nips or charges at you, scold him so that he learns not to be aggressive with you.

You can also make sure his flock has plenty to eat. A hungry rooster is an angry rooster, but if he's well-fed and so are his hens then he will be social and happy with you.

It is not guaranteed to work: Some roosters just have an attitude problem. You can either look at it as a boon to the flock (sure, you can't go near the hens, but neither can predators!) or you can look at rehoming or culling the rooster and finding a gentler guardian for your hens.

With the right steps, a rooster can make an amazing (but not necessary) addition to any backyard flock.

Medicated or Clean Feed

When you go to an online retailer or an in-person farm supply store, you may note that there are two types of chick-grower and layer feed: Some have medicine and some do not.

There are other differences, too: You can find calcium-fortified layer feed or purchase separate crushed oyster shells to make sure your hens have nice strong building blocks for their eggs. You can buy all kinds of snack foods and pellets versus crushed foods versus scratch. It can be overwhelming to choose between all these possibilities. Add to that the complication of medicated feed and what it does, and a trip to the feed store can leave you wondering if you're doing the right thing for your flock.

Chapter 7 will help address the issues of which feeds to use when, but for now let's take a moment to examine the risks and benefits of medicated chicken feed (B. E., 2019).

First off, what is it? Medicated feed contains Amprolium, a medicine designed to combat a condition called *coccidiosis*. It is especially critical during the first 18 weeks of life, when chicks' immune systems are still developing. Eating medicated feed at this age can boost immune resistance to the parasite during your flock's youth.

Coccidiosis can be found in your backyard soil, brought by other birds in their stool, or caught from chickens that weren't quarantined. It is a parasite that causes intestinal issues including very loose stools and blood in the stool, and it can be fatal if untreated.

Because the parasite that causes coccidiosis is impossible to avoid—it is in the soil, in the resident fowl of your area, in the world—medicating the chicks' feed will help them build immunity to it so that when they come face-to-face with it their bodies are ready to do battle against the parasite.

Are there any cons to using medicated starter feed? The truth is, not really. Research has shown that there is no transference from medication to the eggs, and chicks really only need that immunity boost for the first three weeks of life. After that, non-medicated feed is safe for them. There are treatments available if an older group of chickens develops coccidiosis, and immunity is lifelong once a chicken has had the disease.

The good news is that if you have misgivings, the medicated feed is not necessary. Avoiding it is a risk: There's always the chance your whole set of chicks is wiped out in a matter of hours by something medicated feed could have prevented, but the odds of that are reasonably low enough that if medicated feed is not in your comfort zone, it is not necessary.

Culling or Rehoming

Let's say, for whatever reason, you end up with extra roosters or with a rooster that just won't stop with the aggression no matter what you try.

What do you do next?

This is a very personal choice. In social media groups for backyard chicken farming, people can get quite hostile with one another: Some value the rooster's life over anything else, and others see the roosters as meat birds whose time has come.

You have an array of possible choices here, depending on your moral compass when it comes to animals. Remember that none of them are wrong if they are right for you.

It is entirely your call how thorough an investigation you want to do of anyone purchasing or adopting your roosters. You may be curious about what the person plans to do with your roosters: Do they run a fighting ring? Do they cull? Is it a rooster sanctuary? Are they looking for a rooster for their own flock?

Some people need to know the answers to all those questions. They want to ensure the roosters go to a good, loving home, or they want to ensure there is no cockfighting involved. While cockfighting is illegal, it is still found in most states. It may be difficult to discover the truth about someone who is practicing. Look for someone who adopts new roosters frequently, or isn't willing to share pictures of housing.

People can lie, however, and ultimately it is out of your hands if someone uses your roosters for fighting. It is one of the tragedies of the chicken farming world.

Ideally, if you have to rehome you can find a rooster sanctuary to take them, or another backyard flock looking for a rooster to guard their chickens. If you *know* you have an aggressive rooster, you may want to advise an adopting family of that.

Another option, if you aren't open to rehoming or culling, is to create your own rooster sanctuary. You'd need plenty of space for them and a big enough yard that you can keep them extremely separate from any hens.

Whatever you decide, the truth is that if you choose to have one pet rooster with your flock, you will likely end up deciding the fates of many others in the years to come. The same goes for if you choose to incubate your own eggs or purchase straight run chicks from the store.

Knowing where you stand on this matter ahead of time will help you to enjoy your flock with peace of mind that you have a plan in place for any spare roosters that come your way.

Quirky Eggs

One of the more shocking finds for the inexperienced backyard chicken farmer is the inconsistency of eggs. When we buy eggs from the local store or even from local farmers, we get consistency: Hard-shell, white with the yolk inside. Everything looks good!

When you farm your own chicks, you may find more variety among the eggs you discover in your coop (Mormino, 2017).

To understand what goes wrong, it is first important to know what happens when egg-laying goes right.

1. The chicken's ovary releases the egg, what we know of as the yolk. This process takes about 15 minutes.
2. The white of the egg forms around the yolk. This process takes about four hours. The white is made of albumin and the egg at this point is held together by a loose outer membrane.

3. The shell forms. This takes anywhere upwards of 21 hours to complete. Chickens have an actual shell gland that takes care of forming the hardened shell.
4. 20 hours is for the shell formation
5. One hour is for the shell coloring (*not* true for Ameraucana chickens, who produce true blue eggs, not 'painted' blue eggs)
6. The chicken lays the egg in the nesting box.

The two biggest factors, besides disease, that influence eggs are light and access to good calcium.

Some people choose to have a light inside the coop to simulate sunlight for the chickens during the shorter winter days, to push the chickens to continue laying through winter. This method works, but remember that in breeds that typically take a seasonal break from laying, using false light will deny their bodies the recovery period they need and may even shorten the number of years they lay.

Common egg problems and their causes:

- **Thin shell**:
- If your chickens are laying thin shells, the best solution for this is to add a calcium-fortified

feed to their diet. Better yet, add crushed oyster shells.

- **Soft shell**:
- This condition is *true* lash egg, although you may hear that term used to describe a different condition.
- Add calcium to the diet to help resolve this problem
- **Caseous exudate**:
- This condition is *false* lash egg, termed "lash egg" because of a combination of ease of saying "lash egg" and confusion among chicken farmers of the difference between a soft-shelled egg and this mess
- This is a rubbery lump of materials, including some bits of egg and other things that look genuinely disgusting
- It is a sign of an infection called *salpingitis,* in the chicken's reproductive tract
- most often caused by bacteria
- occasionally viral
- can be treated with antibiotics from a vet if it is bacterial
- More likely among high-production hens than hens that lay at a slower rate
- **Coloring troubles**:
- Most eggs get 'painted' inside the chicken.

> Sometimes the spray-painter acts up and you end up with some interesting patterns on your egg. These are harmless.

HAPPY COOP, HAPPY FARMER

When you want healthy, happy chickens, keeping a clean coop is a critical part of achieving your goal. Here are some helpful tips and tricks that help you and your chickens have a good dynamic together (A.T., 2013).

Waterers

While your chick waterer might have been a free-standing inverted bucket with an open base for chicks to drink from, these can get messy if used in a coop for adult chickens.

You can use the original made-for-chicks waterer (or a larger version of the same thing) but hang it from a shepherd's crook or other devices inside your coop, to prevent water contamination.

Trough waterers offer the option of a screen-covered tray that is elevated off the ground to prevent bio debris from getting inside. For use with trough waterers or small-cup waterers, farm supply stores sell special water drip dishes which can be hooked to a PVC pipe system to keep the water clean from contamination. While this type of system takes effort to set up, it helps avoid contaminants in the water and may be worth the extra energy.

If you set up your coop right and live in a rain-friendly climate, you can collect water for your flock in a rain barrel which feeds directly into their water supply. This eliminates the need to have hose access near the coop.

Dust Baths

There are few things chickens love more than a good dust bath. You can have several in your yard so the chickens can bathe in their coop or in the run.

You can fill the dust baths with dried, low-dust sand mixed either with ashes from a fire or with a product called diatomaceous earth, which is a powdery white substance derived from silicone-heavy sedimentary rocks, which chickens enjoy bathing in. Diatomaceous earth helps stop the spread of parasites and keep mites, lice, and fleas from becoming a problem in your flock.

If your chickens are free-range, you can let them seek out and make their own dust baths in your yard. They will put them where they want them, but keep in mind that they may choose your flower garden or a chunk of your lawn under their favorite tree, and the area will never be the same again so long as you have a flock.

Odors

A common but easy-to-tackle problem that crops up in chicken coops is odors. Chickens don't litter box train: They go where they want, when they want, often without seeming to notice. This means that their coop needs bedding on the floor to facilitate easy cleaning, and it needs to be cleaned regularly.

The nice thing is that cleaning is not all you can do to help limit smells in your coop. There are a number of chicken-friendly odor eliminators on the market, which can be purchased at your local farm supply store and sprinkled around the coop after you clean. These will keep the air smelling fresh and clear.

Beyond that, remember to clean the coop out every few days, not only to reduce odors but to reduce the risk of disease.

Sickbay

However hard you work to avoid injury and diseases, odds are that you'll have small troubles crop up occasionally among your birds. It is important to have an already-established area where you can quarantine your sick or injured chicken.

Chickens with injuries need to be isolated in order to protect the chicken from attack by the rest of the flock, prevent infection, and keep the chicken in a calm safe environment while it heals.

PHOJANA PHONBUMRUNG Rooster (2020) retrieved from www.pixabay.com

Chickens who are ill need to be isolated to protect the flock from whatever pathogen the sick chicken carries.

A third reason to isolate a chicken is if one of your hens is violently bullying the others. Normal pecking order behaviors are one thing, but if you have a hen that is regularly drawing blood you can isolate her for a few days. When she returns to the flock, she will be at the bottom of the pecking order again as a 'new' chicken. Hopefully that will break the habit.

When you add to your flock, the sickbay is also a useful place to house the younger generation of chicks until they reach maturity. Introducing them to your flock when they are still smaller than the other birds guarantees pecking order issues and can even cause the younger chickens to suffer from malnourishment as the bigger hens push them away from the feed. Add to this that you need to feed young chickens starter feed and mature chickens layer feed, and you have a compelling reason to keep new chicks separate from your main flock for as long as possible.

The quarantine area does not need to be big, but it should be far enough removed from the main coop and run area that disease cannot spread between the two coops.

If you free-range, you'll need to take extra care to ensure that your flock can't get close enough to the isolation coop to spread any diseases between birds. It does not need to be big, but it should be secluded.

Feed

For the first few months of life, chicks will need a chicken starter feed, which can be medicated or not pending your preference and desire to prevent coccidiosis. As you introduce scraps to the chicks' diet you will need to add scratch as well, to keep their crops moving and prevent sour crop.

Once chickens reach maturity, they will need to eat a layer feed, specially formulated with protein and calcium. You can add crushed oyster shells to their diet to increase their calcium intake, and of course you should continue giving grit if your flock does not free-range.

Feed does not have to be purchased from the store: Did you know you can make your own? Many backyard farmers save quite a bit of money by making their own feed. Others find they hit a break-even point trying to find the right feed ingredients at reasonable prices.

If you decide to make your own, some ingredients you might consider using include (but are not limited to):

- Corn
- Peas
- Wheat
- Oats
- Fish Meal
- Kelp
- Aragonite
- Poultry Nutri-Balancer
- Alfalfa

- Barley
- Oyster Shell
- Grit
- Flax Seed
- Yeast
- Crushed Eggshells
- Mineral Salts

Different sources will recommend various percentage balances between those ingredients, but a combination of those will help your chickens get all the protein, calcium, and other minerals and fats they need in order to grow eggs and develop their bodies (in the case of broiler birds).

Feed for chickens includes scraps, which you can feed straight off your table into a dish or even just onto the lawn for the chickens if they are free-range. There are not many foods that chickens won't or shouldn't eat, but as a general rule you should avoid avocados, green potato peels, dried beans, processed foods and sugars, junk food, and chocolate.

Feeding scraps is a way to speed up making compost for your garden, as the chickens will process the food for you and turn it into something you can spread over your garden in no time. The chickens get yummy food and the nutrients they need, and your garden gets the nutrients *it* needs. You can end up in a great cycle of composting, growing, feeding, and repeating.

Dedicated Supplies

Earlier in this guide we talked about the ease of access to the coop, for cleaning and egg collection and not having to travel far in the winter.

One of the biggest helps with ease of access is having dedicated cleaning supplies for your coop. This will help prevent the spread of bacteria from your coop to your home or other animal housing areas in your yard, and it will mean you have all the supplies you need accessible on hand at any given time.

For coop maintenance, you should have on hand a broom, hose, pitchfork, and cleaning sprays. You can also have the odor-reducing sprinkles if you desire, and buckets or a wheelbarrow for removing soiled bedding from the coop. Another strategy for that would be to have the compost bin or pile located just outside the coop so that when you clean you don't have to transport the soiled bedding anywhere.

Composting

This is one of the most fun parts of backyard chicken farming, if you choose to put in the effort. You can find yourself in an endless cycle of reusing the same nutrients around your yard, to feed yourself, your garden, and your chickens.

It is a fascinating way to learn about science and the nutrient cycle among living organisms—a fun topic if you homeschool your kids or even just have curious littles bursting with questions about how your backyard farm is sustainable.

Keep in mind that too much chicken refuse applied too quickly can damage your soil. It is better to compost the soiled bedding in a bin or pile for a few weeks and then, when it is soft like fresh soil, apply it to your garden. This will keep your plants growing healthy and strong.

The plants will keep you and your chickens growing healthy and strong, and look at that: You have a sustainable yard.

Mealworms

It is a truth universally acknowledged that a chicken in possession of a large tray of layer feed must be in want of a mealworm. The majority of chickens love mealworms, in large quantities when possible. Just the sound of the bag opening can be enough to bring your flock running full tilt across the lawn toward you.

Depending on the size of your flock, providing mealworms as a treat can be prohibitively expensive. If that is the case, don't worry: There is an easy, cheap way to raise your own mealworms.

All you need is a bin or wood box (not cedar), wheat bran or rolled oats (two to three inches across the bottom of the bin), a shipment of live worms, access to water, and a nice warm place for the worms to grow (above 80 degrees Fahrenheit is ideal).

With this setup, you'll have your own harvest of mealworms every two or three weeks, and you can control the size of your mealworm crop. As a bonus, the worms are alive, which increases appeal to the chickens (though that may not seem believable). You'll even have living worms in winter, which birds will be hard-pressed to find in your yard even if they free-range.

COMPANION ANIMALS

Sometimes, for whatever reason, you decide your backyard farm needs more than just chickens. This is a topic that comes up frequently in internet forums for backyard farmers: "Can ____ live with my chickens?" (Griffler, 2020)

Adina Voicu Cat Ducks and Chicken (2019) retrieved from www.pixabay.com

Rabbits

Rabbits are another animal you can keep either as a pet or raise for meat. Some people choose to keep both rabbits and chickens to increase the variety of their sustainable backyard.

Because chickens carry salmonella which rabbits can be vulnerable to, housing them in a shared run is not quite ideal but it tends to work surprisingly well. If you can keep your coop clean enough and keep the chickens' refuse away from the rabbits' feed, the animals tend to graze and forage together like the best of friends.

Llamas

Instead of a rooster, if you have the space for it you should consider getting a llama. Among the barnyard animals, they tend fairly low-maintenance and they are *phenomenal* at protecting your yard from predators, night and day.

Llamas have a whole other set of needs and live for a long time, so you should not undertake a llama purchase or adoption lightly. That said, they are worth considering because they will bond to you and your flock and protect your family better than any guard dog.

Large Birds (Ostriches, Peacocks, etc.)

These can be a great and fun addition to your yard, but be warned that these birds have much stronger odors than chickens. They will also need to be housed separately from your flock and care should be taken to prevent transmission of diseases and parasites between the birds.

Ducks

Like with the large birds, ducks can be fun but should be housed separately. You'll also want to take care to ensure that no male ducks can ever accost your hens, because their anatomies can injure a hen.

Another word of caution about ducks is that they are more likely than chickens to catch diseases from wild waterfowl, and then share those diseases around your yard if you are not careful. (For example, you should not pet a duck and then pet a hen without first washing your hands.)

Pigs

Thomas B. Pig Duck and Chicken (2019) retrieved from www.pixabay.com

Pigs, despite their reputation for being messy, are actually clean and intelligent animals. They tend to create their own pecking order and want to be the alpha 'kid' in your family (they can and will inform your children who is boss).

Chickens will groom pigs, keeping bugs and external parasites away from the pigs. In the same vein, pigs constantly root and turn over the soil. A smart companion chicken, following behind the pig, can then forage for grubs much more thoroughly than if the bird had to turn the soil itself.

Pigs do not have a lot of natural predators, and if you keep chickens with your pigs the mere presence of the sows will be enough to deter most classic chicken predators.

The biggest issue with keeping pigs and chickens is that the pigs, true to their names, are likely to go after the chicken feed and the eggs. They may even try for a live chicken if it gets in the way at feeding time!

Goats

Goats and chickens do fine grazing together but they need separate housing at night. During the day, free-ranging, the goats will protect your flock from predators and the chickens will protect your goats from flies and other insects and external parasites.

So long as they go home to their own houses at night, goats and chickens should be able to free-range together without difficulty.

Barn Cats

Cats are, in their nature, predators. They are also, fortunately, very lazy creatures and they will usually leave your birds alone once they're accustomed to the idea of a flock. Cats are intelligent enough to know that the chickens matter to your family and in most cases that is enough to deter the cat from ever hurting a chicken.

Chickens should be kept away from newborn kittens, as a chicken may see them as food.

Turkeys

Chickens and turkeys do well together in a shared run. Turkey toms will generally leave your hens alone so there is little to no risk of injury from the toms. The turkeys will help guard your flock against predators.

5

BRINGING UP BABIES

STARTING YOUR FLOCK

Lolame Two Chicks (2020) retrieved from www.pixabay.com

When it comes to acquiring your flock, you have a myriad of options to choose from (Backyard Chickens, 2020).

- Purchase live chicks from a farm supply store
- Order live chicks online from a hatchery, with direct airmail
- Order fertilized eggs online from a hatchery
- Purchase live chicks from a local chicken farmer or hatchery
- Purchase fertilized eggs from a local chicken farmer or hatchery
- Buy adult chickens from a local chicken farmer or hatchery

Each option has perks and drawbacks, which you should be aware of as you do your research. You are so close to bringing your flock home, your coop is ready, your heart is ready. It is almost time to welcome your flock.

Some vocabulary to take into account as you research your options:

- **Pullet**: A young female chicken
- **Cockerel**: A young male chicken

- **Straight Run**: Young chicks whose gender is unknown
- **Bantam**: A type of chicken that is ½ to ⅔ the size of a regular chicken. Their eggs are smaller (you need three eggs for every one regular chicken egg, in cooking). If you want a larger flock, bantams will give you more animals in the same space. If you want meat or eggs, bantams are not the best choice.

Farm Supply Store

This is possibly one of the easiest ways to get chicks: You go to the store, you point to the ones you want, and you bring them home in a little box.

Experiences with farm supply stores can vary greatly. Some stores have reputations for correctly-labeled chicks, well-cleaned bins for the chicks to live in, dry living conditions, and healthy birds. Others have reputations for anything but. It is important, if you purchase from a farm supply store, to research the store's chick sales history and look for reviews online.

Another factor to consider, when buying from the supply store, is that they get all their chicks from hatcheries. There may be an upcharge for the service of ordering, shipping, and tending them: You may be able to find the same chicks for a portion of the price by ordering directly from the same hatchery yourself. This may vary from store to store, so again it is important to do your research.

Farm supply stores will label their bins according to bantam, pullet, and straight run, and (with the exception of bantams) they tend to label carefully by breed.

You'll want to isolate any chicks you bring home from other chicks and chickens you have. A general rule of thumb is to isolate and quarantine new animals for four weeks before introducing them to other groups of animals or attempting to add them to your main flock.

If you already have older chickens, chicks should be about 18 weeks old before you attempt to introduce them to the older flock.

Hatchery Chicks

Another option is to purchase mail-order chicks directly from a hatchery. Generally the hatchery will give you an estimated shipping date, but because they have to ship as soon as the chicks are fully pipped, this date may be adjusted depending on hatch rates the day your chicks are supposed to ship.

The way hatcheries work is important to be aware of:

1. Chicks hatch. The number per breed per day

can vary unpredictably, which may impact your shipping date. Consider it more of an estimate than a guarantee.

2. When chicks first hatch, they take time to absorb nutrients from their eggs. This is why hatcheries are allowed to ship live chicks on the first day, because they do not need feed or water to survive in the mail. All they need is ventilation.
3. Your chicks will be shipped to you as quickly as possible. This may come with a hefty shipping fee.
4. When your chicks arrive at the post office, expect a phone call to come pick them up immediately—even if it's the middle of the night.
5. Sometimes chicks do not survive transport. Because of this, some hatcheries might send extras. Others might send replacements.
6. Once you have the chicks you can establish them in your brooder set up and get them settled. (See steps for that later in this chapter.)

Hatchery Eggs

This option is to order fertilized eggs through a hatchery and have them delivered by mail. It comes with mixed reviews online and seems to vary less by hatchery and more by luck of the draw. Incubating eggs (see steps for egg incubation later in this chapter) is full of opportunities for missteps. Even perfect incubations can result in no hatched eggs.

Still, it is a way to find more rare breeds and, since many hatcheries will only ship a minimum quantity of live chicks, it is a way to have more control over quantity. Just because you order a dozen eggs, typically for much cheaper than the cost of live pullets and cockerels, doesn't mean you will *hatch* a dozen eggs in your incubator.

Another line of worry is that there is some belief, without research to back up the claim, that eggs shipped over airplanes have chemical changes take place due to the changes in pressure while on the plane. The idea is that eggs flown by airmail may be less viable than eggs shipped over the ground. While there is no evidence to support this notion, outside of anecdote, it may be worth finding local hatcheries that can ship ground only to avoid the risk.

Live, Local, and in Your Backyard

If you choose to visit either a backyard chicken farmer, an industrial chicken farmer, or a hatchery, you can see that the chicks are healthy and alive. Often there is no minimum purchase amount—you can leave with just one very special chick or with a hundred if your heart desires.

This method has the advantage that your chicks do not endure any hardships while being transported in the mail. You can examine chickens before you buy them, for any disease or congenital troubles.

When you order from a hatchery, frequently they have a minimum order size, but when you shop in person you can select just one of each breed you want, or any permutation of breeds and quantities.

You may also make friends and personal connections in the chicken farming world, as well as supporting local farmers. When you have trouble with chicks you've purchased locally, you may be able to reach out to the farmers to see if they've had similar experiences or if they have any advice.

Overall this is a good method to buy chicks, with some hidden chicken networking benefits.

Dormant, Local, and in Your Backyard

Like buying live chicks from a local place, buying fertilized eggs has some clear advantages, especially when it comes to supporting local farmers, establishing connections, and reducing stress to the eggs. You can often find local fertilized eggs for fairly cheap.

Remember that when you incubate your own you are getting straight run chicks: You are all but guaranteed to hatch roosters. Please keep this in mind as you plan your flock.

That said, the nice thing about hatching your own is that there is no quarantine process. The chicks don't catch anything from their eggs, so they are born healthy and ready to meet any other chicks you have.

Further, if you have an established flock with any broody hens, you can sneak the eggs (or even the new chicks!) under the hen and let her do the work. While that may not be easy this time if you're just starting out, it is absolutely an option for every other egg order in your future!

Some breeds—most notably Wyandottes—are great broody hens, so keep that in mind if this is an option you think you'd like to pursue.

Bigger Friends

It's possible you don't want the wait or the work of raising hens from pullets: You may want the eggs now, from healthy adult birds that are ready to be at home in your yard.

Many, many hatcheries and industrial chicken farms offer this option. You will pay more for the hens, as they've cost the farmer feed, effort, and time, but what you typically get is an established, healthy flock.

Sometimes, especially if you choose to buy from an industrial farmer, the chickens may have been cooped up in a too-small area. This means you are rescuing them and giving them a better life, which is so good for them, but it also means that they will likely come to you stressed, short quite a few feathers, and possibly needing to learn how to forage.

If you bring these bigger chickens home, they may start laying immediately, but there's a chance they will need a few days to adjust to their new home before they start laying.

Because these older hens did not grow up knowing you, you'll want to get some feed or treats—mealworms are a good option beloved by most (but not all) chickens—and work to train the chickens to eat from your hand and trust you. You can even get them to come when you call them, in time!

INCUBATION OR BROODY HENS

Sometimes the decision to welcome new chicks into your flock isn't up to you: The hens may decide it for themselves (Backyard Chickens 2020).

Whether you've ordered or gathered eggs with the intent of incubating them, or whether you've been surprised to find broody hen on a nest one morning, some fundamentals are consistent across the board when it comes to growing and hatching a new set of chicks.

The incubation process itself is fraught with idiosyncrasies and quirks that can make it frustrating to try to hatch any living eggs. Before you commit, you should know that the process takes an average of 21 days between the first day you grow the chick to the first day you can hold it in your hand. The 21-day average can range from 18 to 25 days, but eggs that still have not hatched after 25 days should be discarded.

Eggs have to be kept constantly warm. When you incubate, this means ensuring that the machine is working right and regulating temperature well. When you have a broody hen, she will do all of that work for you.

When the incubation period is over, things become more emotional. The process of a chick breaking out of its shell is called *pipping*. It can be thrilling to watch, and it may be tempting to help struggling chicks out with this process, but there are numerous reasons not to.

- Helping them could introduce infection.

- If they are not pipping on their own, there is likely a medical reason.
- If you help weaker chicks hatch and then breed them into your flock, *their* chicks will be even weaker/worse at pipping and have more health problems.
- Humidity (in an incubator) is critical—if you open the incubator to help some chicks pip, you will likely kill others that would have been fine.
- If a broody hen has stopped sitting after some chicks hatch, she has decided the remainder of the eggs won't pip. Her instincts are likely right.

Incubation Process

Before you put eggs in your incubator, the first thing you will want to do is take it for a dry run. Plug it in, get it started, check the temperature and humidity several times a day. Find out what quirks your incubator is prone to.

Each incubator, even between two incubators of the same make and model, has its quirks. Since you are counting on your incubator to grow the lives that will help to feed your family, understanding your incubator's specific idiosyncrasies is critical.

Once you feel you understand the needs of the incubator, how often to check and regulate the temperature, how often to add water, etc., it is time to add the eggs. Each incubator will come with specific instructions for how to do so: Some have individual slots for eggs, others have levels between plastic grills that slide when it is time to turn the eggs. Be sure to read all instructions specific to your incubator before adding the eggs.

Before you add any eggs, you can use a tool called an egg candler, to check for fertility. This special light will allow you to look inside the eggs for blood vessels that indicate a chick is developing. Candling becomes much easier after three days of incubation. From here you can tell which eggs are progressing and which are not.

You should turn the eggs a minimum of three times daily (every eight hours, to the best of your ability), or up to five times daily if you can manage it. Within the range of three to five turns, more is better but not so critical that you should panic about it if you have a busy day now and then.

For days 1-18, your job is to check the temperature and humidity, regulate them as needed, and turn the eggs up to five times a day. For simplicity, it may be easiest to check temperature and humidity when you go to turn the eggs.

Once you hit day 18, things change: Your job is to leave the chickens alone. They go into what is colloquially referred to as 'lockdown.' You should still check on temperature and humidity, but you should not turn the eggs. The chickens don't need it: They're busy at this point moving into position to start pipping and turning them will prolong that process.

This lockdown period can be emotionally taxing: The temptation to help the chicks pip is strong, and you have no way to know which chicks will and won't hatch. Out of a dozen, you could get none or you could get all twelve! It is important not to interfere, whatever your temptation might be.

During and immediately after pipping, the chicks will need time alone in the incubator to absorb their membrane, become fluffy and dry, and acclimate to life —they need to adjust to seeing and hearing the rest of their flock.

Once pipping seems to be done, and all chicks are fluffy and dry, you can open the incubator and move the chicks to their brooder. You *can* at this point, perform a float test on the eggs that haven't pipped:

- Make a dish of deep, *warm* water
- Set an egg in it at a time, without any other motion around
- If the egg rocks, the chick is still alive
- You can help it pip at this point (use a candler to

find where the air bubble is and start at that end)

- There is no guarantee of health for chicks that are unable to pip on their own
- Any eggs that do not rock will not pip and can be thrown out

When you move chicks to the brooder, dip their beaks in the water to teach them how to drink. They will likely find the food on their own, but the quest for water in new chicks is important and you can teach them how.

Broody Process

The overall concept of brooding is similar to incubation in terms of the number of days, but the type and amount of work you put into it are different.

The first and most notable difference is that your primary job is to take care of the broody hen. She will likely not leave the nest for more than an hour a day. Make sure your broody hen has access to food and water without moving too far from the nest. Give her protection and seclusion from the rest of the flock, and ensure she is in a secluded place where she can relax and do her job.

Like with an incubator, the chicks need 21 days, or

three full weeks, to go from freshly-laid eggs to hatching. Also as with an incubator, there is no guarantee of chick quantity: The broody hen could sit for three weeks and not hatch any, or she could hatch a flock bigger than the one you already have.

While you don't want to disturb her, having a count of the eggs she's sitting on (you can check this during her hour or so away from the nest) will help you gauge the best— and worst-case scenarios. Some chicken farmers label the eggs the broody is sitting on with a pen, so they can remove any extras she adds over time.

You can time a new arrival of hatchery-bought chicks for around the 21 days, if possible: It's an easy way to introduce new chicks to your flock and ensure they'll be accepted and protected by the hen. The best way to do this is to go into the coop at night and tuck the living chicks under the hen's feathers while she sleeps. She'll wake in the morning, proudly welcome her "new babies" and they'll be accepted by the flock. This is a convenient way to ensure that you only introduce pullets and don't have competing roosters to contend with. You can even do this with broody hens when you don't have a rooster to fertilize her eggs. She'll feel accomplished and you'll have a bigger flock without the usual pecking-order issues that go with introducing new hens.

Whether you add chicks to the mix or not, the hen is sitting on eggs for three weeks. Once those start to pip, she may become distracted by the already-born ones and abandon the remainder. Just quietly tuck them back under her whenever she's sitting on the newborn chicks. Any eggs remaining on Day 25 should be discarded.

Myriam Zilles Pipping Chick (2020) retrieved from www.pixabay.com

With an incubator, you need to watch the newborn chicks to make sure they absorb the membranes and dry out well, but a broody hen will take care of all of that for you, and give your new chicks a brooder. In fact, if you hatch or adopt out chicks to a broody hen, you can ignore the following section on how to set up your brooder.

BROODER BOX

For anywhere from the first six to the first eighteen weeks of their lives (depending on how you choose to do this), the brooder box is where your baby chicks will have their early experiences. This is where they will learn about each other and you, about the world, and how to forage and eat. This is where their health will be established (Backyard Chickens, 2020).

The First Two Weeks

During the first week, the temperature in the brooder should be as close to 95 degrees as possible. Some people go as much as 105 for the first few days, but if you're going to have a heat source or lamp running that hot there should be room within the brooder for chicks to get away from it.

There are some simple guidelines to follow so you can monitor how your specific chicks are faring in your brooder. Just because a book says something about temperature doesn't mean your chicks will be happy in that setting.

If the chicks are happy and evenly-dispersed throughout the brooder, the temperature is just right. You want to check for this at least twice a day to prevent coldness and dehydration in your chicks.

If the chicks are huddled under the heat source, the temperature is too cold and the lamp should be lowered or the heat source increased.

If the chicks are as far from the lamp or heat source as possible within the brooder, it is too hot. The lamp should be raised or the heat source reduced.

If your chicks seem lethargic, there could be a number of causes but it is worth checking to make sure the brooder isn't simply too hot, before you panic about other medical conditions. A quick fix like a temperature change may save a chick's life.

Your brooder should have substrate on the ground. Some people make the mistake of trying to keep the chicks in a cardboard box for the first few days, which they can then dispose of, but the flatness of the cardboard can hurt little chick feet.

Because chicks tend to peck at the ground and eat anything they can get into their mouths, it is important to choose a bedding that is safe for them. Cedar and paper-based beddings sold at some stores are not ideal for chicks (or older chickens). Your best bet is to invest in pine chips or hay.

Your brooder should include a waterer that you can clean and change out at least once daily. Some people add marbles to the brooder water to keep chicks safe from drowning risk. Adding apple cider vinegar in the first few days can help boost chick health and prevent pasty butt (we will cover pasty butt in a few paragraphs).

Chicks should *not* have electrolytes in the water on their first day in the brooder if they were mail-ordered chicks. They need time to hydrate before electrolytes are introduced to their diet. After the first day, however, including an electrolyte supplement to the water can be beneficial to chick health.

The feeder you use in your brooder can be fairly utilitarian, so long as it can be cleaned routinely. Chicks have a charming habit of kicking their bedding into the feeder, so be prepared to repeatedly clear the feeder of any bedding materials. Remember at this age to use chick starter feed and consider using the medicated variety to help build immunity against coccidiosis.

Another thing your brooder should include is a secure covered area for the chicks to sleep under. This will mimic the broody hen, and it can be something as basic as an overturned cardboard box with a wide opening, or it can be something more creative like a hanging feather duster to simulate the body of a broody hen.

Whatever you choose to use for the covered area, this should be located away from your lamp or other heat source to reduce the risk of fire.

Taking these steps will ensure your chicks are warm, watered, and fed, but what other issues do you need to forestall if possible? Earlier the phrase *pasty butt* came up. What is it, and what do you do for it?

Chicks sometimes have trouble fully emptying their vent when they do a bowel movement. If their waste dries against their vent, it can solidify inside them and prevent them from having any further bowel movements, which can prove fatal. During the first week of their life, you'll want to check your chicks for any vent trouble a couple of times a day. If there is any stool build-up, you can clean it off with a damp cloth.

You do not need to fully remove it from their fluff if you have trouble getting it soft enough to clean off. The other chicks will help keep the flock clean. All you need to do is ensure that the vent itself is clear of any obstruction and can flow freely. After the first week, pasty-butt is no longer a concern.

Two Weeks and Older

After the first two weeks, you can move the lamp or adjust the heat source to lower the temperature by 5 degrees Fahrenheit per week of age. So, at two weeks old, the temperature can be lowered to 90 degrees Fahrenheit, at three weeks it should be 85, and so on.

After two weeks, you can start introducing food scraps to your chicks—chopped up pieces of salad, scrambled eggs, leftover dinner food. Remember that chickens shouldn't have potato skins, avocados, chocolate, sugary foods, dry beans, citrus, or anything moldy/rotten (moldy foods can trigger sour crop, which can be fatal—it is not worth the risk).

Beyond that list, your chickens can eat just about anything. For young chicks, you will want to chop it up nice and small, to about the same size as chick starter feed, so that they can eat it easily.

Once you add scraps to the chicks' diet, you will want to add something called *scratch,* which helps the chicks process food in their crop.

Many chicks also love mealworms, which you can buy in fairly large bags from your local farm supply store, order online, or even grow yourself in your backyard to save money. Scratch will help the chicks digest mealworms as well.

Some backyard farmers find at some time between

three and six weeks, their chicks start to outgrow the space in the brooder. By six weeks, so long as your chicks are fully-feathered, active, and healthy, they can be moved outdoors. Depending on the season and the weather, you may need to keep the heat source available for them at night in the coop.

Remember during this period—the first six weeks and then the following twelve as well—the more you handle your flock, the more social they will feel toward you. This is especially important when you have a rooster, as some unsocialized (and, let's be fully honest, some perfectly-socialized) roosters can become aggressive toward people.

By eighteen weeks, your baby chicks are full-grown chickens and your hens may even be laying by now. If not, they will be soon.

6

DAILY LIFE IN THE BACKYARD FARM

A boon to your backyard farm will be routine. If you can get down a system that works for you, your chickens will know what to expect and you will avoid falling into any bad habits (Chicken Keeper, 2020).

Part of the routine is personal preference: When and how do you want your flock to wake in the morning? When and how do you want them to eat, lay eggs, and go to bed at night?

They seem like subtle decisions, but everything from where you put your coop to the shape of your yard can influence what kind of routine you have.

The first thing you should do is take a look at your

existing life routine. You ideally want chickens to integrate into this with ease so that you have as little disruption as possible. Backyard farming, though it is added work to your day, should ultimately be a fun and rewarding experience.

Consider your needs: What time do you go to work? Do you have kids who need to meet a bus, go to after school activities? Can you delegate any or all of the chicken chores?

There are ways to mix the multiple goals of bonding with your chickens, teaching your children responsibility (if that is a factor in your life), and having a happy backyard chicken flock.

DAILY TASKS YOU CAN DO ANY TIME

Depending on your schedule, you may prefer to do certain tasks in the morning before you start the rest of your day, or closer to the end of the day. These are a list of daily chicken care tasks that you can assign to any schedule you like, so long as they get done:

You'll want to check your feeders to ensure they have enough food and there is no contamination by bugs or rodents. This is one of the best reasons to design a coop that is off the ground: Rodents have a more difficult time infiltrating feed bins they can't dig their way into. This daily check on the feed is necessary even if you use an automatic feeder.

You'll also need to change out the waterer every day. Wash it with dish soap and hot water to thoroughly clean and disinfect it before adding fresh water and returning it to your coop or run. This daily task helps to prevent parasites and bacteria and will assist with keeping your flock in good health. As with the feeder, this task is necessary even if you use an automatic waterer or collect water in a rain barrel for the chickens' use.

If you have indoor and outdoor feeders and waterers for increased chicken access day and night, be sure to check and change both. Changing an outdoor waterer is *critical* for your flock because your local indigenous birds may use the water and spread diseases to your flock if it is not kept clean.

MORNING TASKS

Some chicken upkeep tasks must be done in the morning, for fairly self-explanatory reasons.

As a starting point to their day, the chickens need to be let out of the coop, into the run or the yard if they are free-range. Some people choose to install automatic door openers (many of these are programmable to open at dawn), but you should be aware of a possible drawback to these systems: Predators can learn about these doors and, rather than waste hours of their night digging to gain access to the coop, simply wait by the door to strike at your flock as soon as it opens.

If you decide to have automatic door openers, they are best paired with a security system (described later in this chapter).

At this point, you should feed your chickens: This can be a mix of scraps and layer feed. Chickens should be given access to water before food, to get their crops moving. Do not spread feed on the bare ground, as this can increase the risk that they catch a parasite or bacteria.

If you have an aggressive flock, you may need to supervise the morning feed to ensure that all chickens get access to a healthy amount of food. This is also a good opportunity to hand-feed the chickens any treats so that they learn not only to trust you but to look to you for good things.

You'll want to check the coop for eggs. This makes a great morning routine because doing the egg check at a set time daily ensures that you know the eggs are fresh.

You'll notice any egg-bound or broody hens right away, as well as any changes in your hens' laying habits.

Remember that if you encounter a broody hen, you'll need to provide them with separate water and feed they can access without leaving the nest. If possible, give them a secluded area to brood. You can mark their eggs with an x or with the date. Marking the eggs on the first day of brooding will ensure you know exactly when to expect chicks. It will also prevent her from adding new eggs to her brood pile after she starts, which prevents wasted eggs.

This time of day with your flock can be a special bonding period for you, or you can delegate it to your children. Feeding and tending the flock, checking for eggs, reporting anything unusual to you: All of these activities promote responsibility in the growing mind.

AFTERNOON TASKS

In the afternoon, you or your children do not need to do much unless you want to take this time to bond with your flock a little.

You'll want to top off any feeders or waterers that are getting low, but you do not need to clean the waterers more than once per day.

Take this time to check on your flock. If any chicks were egg-bound this morning, how are they doing now? Check-in on any birds in quarantine, and tend any injuries among your flock.

Bonding can be achieved by sitting near your flock, feeding them by hand, delivering special treats, even by talking to them. Flocks usually have one or two outgoing birds that may choose to sit on your lap.

Taking time to let your flock learn to trust you may feel like a waste if you look at your chickens as little more than layers and broilers, but keep in mind that this can make it easy for you to treat wounds and injuries as they come up because your birds will know you are there to help them.

EVENING TASKS

In the evening, close to dusk, you'll want to put away any uneaten feed. This keeps night animals from foraging in your run and reduces the risk of attracting predators. Keeping food locked away at night will prevent a pest infestation in the food and saves you money by reusing today's feed tomorrow.

Once the food is put away, you'll want to secure the chickens in their coop. A complicated latch or a padlock will help guarantee that predators cannot get in during the overnight hours.

As in the morning, there are automatic door closers you can program to shut at dusk. If you use these, you'll want to first ensure that your birds always go in the coop independently before dusk, and you'll likely want a security camera so you can ensure that no predators get inside the coop with your flock.

SECURITY OPTIONS

With the advent of modern security technology and the incorporation of smartphones and security, there is one option backyard chicken farmers didn't have even a decade ago: You can set up motion-detection surveillance around your coop.

With this technology, you can:

- Be aware of which predators are casing your coop
- Check on your flock remotely from work during the day
- Be precisely aware of which predators got to your flock if you have an incident
- Have peace of mind about your birds

Such security systems are becoming increasingly affordable and adjustable to your specific needs. A coop camera will also make it so you can safely use the automated doors, because you'll know if predators wait for the door to slide open in the morning.

7

HEALTHY, HAPPY CHICKENS

Most of the breeds mentioned in this guide book are resilient and not susceptible to diseases. Still, there are some conditions and illnesses that any fowl is vulnerable to, and it's important to be aware of these. Prevention and planning can help you avoid serious medical emergencies and fatalities among your flock.

PREVENTION IS THE BEST PROTECTION

There are some basic tips you can follow to help keep diseases, illnesses, and injuries away from your flock. While they aren't a guarantee, they can certainly help protect your backyard chicken farm (Kathy, 2014).

Most critical of all is to ensure that your water supply is

clean and changed daily. Water is essential to all life, including parasites and bacteria, which makes it an easy vector for disease transmission. Keeping the waterer clean and filled with fresh water daily, will help to keep your flock in good health.

You'll want to check the feed daily for any changes in consistency (which indicate it got damp and may mold) or signs of infestation. The lucky thing about chickens, which eat bugs all day, is that if you get a bug infestation in your feed the chickens may just consume the bugs along with the feed. Still, this is not the healthiest option, and bugs can leave behind detritus you do not want your flock consuming.

When you feed scraps, be sure never to give moldy foods. Mold in the crop can trigger a yeast infection which can cause the digestive system to overflow. A bird with a sour crop will lose weight and may die if it does not receive the proper care.

Other foods to avoid giving your birds include avocados, green potato skins, citrus, chocolate, junk food, and foods high in sugars.

You can supplement feed with vitamins and minerals to safeguard their health and be sure you're providing all the nutrients they need to make eggs.

Your birds need plenty of space. As a review, chickens

need ten inches of linear roost bar per bird, three square feet of space inside the coop, and 25 square feet of space inside the run. You don't want to go much over the three square feet per bird inside the coop as this can make it difficult to keep your birds warm in winter.

The coop itself should be well-ventilated to prevent the spread of respiratory issues and to prevent invasive pathogens from taking hold in coop bedding. Try to keep the humidity as low as possible as humid conditions encourage the growth and spread of bacteria.

Whenever possible and recommended for your region, vaccinate your chickens against preventable diseases. This can save you unnecessary stress down the road and can prevent losing all or part of your flock to unnecessary illness.

Know your birds. The more familiar you are with their bodies, their quirks, the earlier you can detect problems in your flock. Daily health checks go a long way in tackling diseases and injuries early, before they become an emergency.

If you have breeds that lay through the winter months, you can have supplemental lighting in your coop to simulate sunlight and help your birds absorb nutrients through the cold season.

Lastly, if possible have an avian veterinarian you trust

with your flock. Many regions do not have avian vets who will treat chickens, so this is often something of a pipe dream, but it brings a certain peace of mind to know you have an expert you can call in times of need.

WHEN PREVENTION IS NOT ENOUGH

It happens eventually, no matter what you do: Something impacts either a single bird or your whole flock. It can be enough to trigger panic and stress. Remember to stay focused on the clinical aspects of helping your birds and try to keep emotions at bay. This will help you help them.

Signs of a Sick Bird

If you are checking your flock daily, you may notice some early signs of trouble. This is ideal: If you notice problems early, you can catch them and treat them before they become a life-threatening emergency or infect your whole flock.

Healthy birds will have shiny feathers. If their wings and feathers look ruffled, dull, half-hearted, you may want to check your bird for signs of injury or infection.

Similarly, if a bird that is usually active and forages all day suddenly becomes lethargic or slows down a lot, poor health may be the culprit. The same thing can be the case if your top-of-the-pecking-order hen suddenly loses her place without explanation.

Other signs that indicate a sick bird include a runny nose, watery eyes, droopy body language, a pale or floppy comb (remember that some chickens do have naturally floppy combs: this is one of many reasons it is so important to know your flock and be familiar with their specific appearance and behaviors), coughing or wheezing, mangy appearance or patchy feathers, and breathing heavily. One of the biggest tips that a bird is struggling to breathe is that they hold their wings out from their bodies to open up their lungs as far as possible.

First Steps

As soon as you know one or more of your chickens is not acting like his or her usual self, there are some preliminary measures you should take to ensure safety for the bird as well as the overall health of the flock.

The absolute first thing you should do is isolate the bird immediately. Earlier in this guide, we discussed the importance of having an established 'sickbay' area of the coop for situations just such as this.

Isolation will keep the other chickens from pecking at the sick or wounded bird, it will prevent the spread of diseases around the flock, and it gives you a chance to observe the bird one-on-one.

(When you are done treating the sick bird, it is important to return to the main coop and do a deep clean so that you get rid of any germs, but you should address the sick chicken's needs first.)

When you have an idea of what you're dealing with, you may or may not need to reach out to your veterinarian. Some conditions can be treated easily at home, and the more experience you gain the easier it will be for you to determine which things need a vet and which just need some tender loving care.

If you decide you need to contact your vet, don't forget one of the biggest advantages of smartphone technology: You can take pictures of your sick bird and send them to the vet. While this won't always tell the vet what is wrong, it can avoid an office visit if the vet can identify the problem over the phone.

Some conditions can only be treated with a prescription, which is why it is so important for you to have a vet available if you can manage it.

COMMON CONDITIONS, SYMPTOMS, AND TREATMENTS

(Weston, 2018)

Coccidiosis

This condition is caused by a parasite that lives in soils around the world. The chickens get it into their systems just by eating grubs or scratch from the infected soil, and the illness wreaks havoc on their digestive systems.

Signs and symptoms of coccidiosis include diarrhea or loose stool, changes in appetite, pale comb, bloody vent, weakness and lethargy, weight loss (or slowed growth in new chicks), and a change in the laying pattern.

Treatment for coccidiosis is a product called *Amprolium,* which stops the parasite from reproducing and perpetuating inside the chicken's body. To administer, you add the Amprolium to the chickens' water for seven days. Even if only one chick is sick, you should treat the entire flock.

Wry Neck

This is a troublesome problem typically caused by poor nutrition in your chickens. In this condition, your bird will have difficulty standing, may be permanently looking upward, or may have a twist or crook in their neck.

So long as the chicken can eat and drink, this is not fatal. It's also not an ideal way for your bird to live his or her life, and it will send them straight to the bottom of the pecking order, so treatment is wise.

You'll want to isolate the bird in question to protect it from the flock and administer a chicken-friendly multivitamin two to three times a day. If the chicken is not able to feed itself, you'll need to take whatever measures are necessary—water with an eyedropper, soft crumbly food—to ensure that your bird is getting the other foods they need.

If one bird in your flock is struggling from wry neck, odds are that other birds may be tending toward malnourishment as well. Take this opportunity to give a day or two of multivitamins to the entire flock, and treat the birds with increased Vitamin E intake for a week or two, to prevent this cropping up elsewhere in your flock.

In the affected bird, symptoms may seem to get worse before they improve, but don't give up: Though wry neck can take up to a month to treat, your bird will emerge healthy and have a special bond to you because you nursed it back to health.

Mites

Mites are one of the many external parasites that can negatively impact the quality of life for poultry. Usually, they are hand-delivered to your flock by rodents or by indigenous birds visiting the coop area.

Mites only live five to seven days on your bird, but in that time they can wreak an incredible amount of havoc. Just one mite is capable of lying up to 100,000 eggs during its life. It lays these on the host bird, but when the eggs hatch they will spread throughout your flock.

Mites bite your bird for sustenance. They draw blood, causing wounds, infection, anemia, and (if untreated long enough) death.

If your flock develops an infestation, one of the first signs may be that your chickens refuse to enter their coop. Since this is their home and their safe place, you should be alarmed any time they won't go inside.

Other signs include lost feathers, red and black spots near the vent, and obvious patches of damaged skin.

You can use a product called *Sevin Dust* to treat for mites, but be aware that its use is controversial because it is not clinically approved for use in poultry and there is no strong scientific research on the topic of egg withdrawal for Sevin Dust.

A dust bath containing diatomaceous earth is a great way to both treat and prevent the mites. During an infestation (and ideally at other times too) your flock should have round-the-clock access to diatomaceous earth.

During the treatment process, you'll want to spray your coop with either a garlic water mixture or a combination of water and dish soap. Clean out bedding before you do this and add fresh bedding once the coop is dry.

Prevent future outbreaks by making sure your flock has routine access to dust baths. If you live in a climate prone to heavy or frequent rains, consider offering a dust bath inside the coop where it can be dry all the time.

Lice

Another external parasite, the idea of lice may make you cringe about a flock infesting your whole family with head lice, but you can rest assured that the lice which plague chickens are distinct from the lice that would cause trouble for your family.

Like mites, lice spread from bird to bird through close contact. The majority of lice spend their 24 days of life on the same host, feasting off their blood, although some lice occasionally move from bird to bird (this is how infestations start). Generally, lice are brought to your flock by other wildlife, especially indigenous birds.

The lice draw blood during their biting phase, which can lead to infection and anemia in your birds.

The nice thing about lice (if something nice must be said) is that you can visibly see them infesting your birds. You will notice nits on the tips of the feathers, and full-grown lice spread throughout the feathers.

Aside from seeing the lice with your own eyes, you may notice that your flock becomes intensely focused on grooming and seems to have their feathers literally ruffled. They may lose feathers and, if the infestation becomes severe enough, they may develop a pale comb and wattle, redness near their vent, and a clear deposit of white eggs on the skin.

You can treat and prevent mites with diatomaceous earth dust baths, spraying the coop down with garlic water or dish soap, and—if those treatments fail—a prescription for pyrethrin spray from your vet should do the trick nicely.

Botflies (also known as Flystrike)

A botfly attack is easily one of the most unpleasant aspects of being a backyard chicken farmer. Hopefully, this never happens to you or any of your birds.

This section gets a bit unpleasant to read through, so feel free to skip over this and only refer to it later if you ever experience a botfly infestation.

Botflies are terrible insects regardless of which animal they infest. They cause one of the most disgusting things you as an animal owner can experience in animals under your care. With a chicken, what they do first is bite their way into your bird. Once they have broken down the skin, they lay eggs in the opening.

These eggs then hatch into maggots, which *eat your chicken alive*. This causes systemic shock and death if it is not treated promptly. A bird suffering from flystrike absolutely ought to see the vet, but there is a treatment process you can follow if you do not have vet access.

It will be devastatingly evident if one of your chickens is enduring this condition: Maggots will be eating their flesh.

That said, there are some warning signs to watch out for, geared toward prevention. For one thing, botflies strike exclusively in warm weather. If the air has been too damp and rainy to allow for routine dust baths, you will want to be on the lookout for a botfly strike.

One of the biggest telltale signs your bird is at risk, however, is if they are already suffering from a medical condition. Frequent loose stools can cause skin breakdown around the vent, and this vulnerable skin is an appealing setting for botflies to lay their eggs.

Treatment is unpleasant for you and the bird but unequivocally necessary if you intend to save the chicken.

Steps of treatment are as follows:

1. Isolate your chicken
2. Feed them plenty of electrolyte-heavy water
3. Bathe chicken (as in, submerge all but the head in water) to drown maggots
4. Use tweezers to remove as many maggots as possible
5. Apply betadine *or* vetericyn spray *or* saline solution to wound
6. do NOT apply peroxide, as it dehydrates flesh and will prevent wound healing
7. Dry the wound area thoroughly—you may need to cut back feathers to do so
8. Continue applying vetericyn spray, but if it does not prevent infection you will need to see a vet for antibiotics

9. Repeat at least twice a day until the wound fully heals
10. do not reintroduce your chicken to the flock until the wound his healed, or they will peck at it

Worms

These are intestinal parasites that invade your chicken —usually through contaminated drinking water—and cause digestive troubles for them. The parasites feed off nutrients the chickens need to survive and lay their eggs inside the chicken.

If your chicken has worms, you may notice weight loss, bloody stools, loose stools, blood in and around the vent, a pale comb and wattles, decreased activity, a reduction in egg production, and...perhaps most visually gross for you... worms *in* the eggs when you go to crack them open.

To prevent a worm infestation in your flock, you must keep the bedding dry and clean. Change the chickens' water daily and wash the container daily. Put feed and water dishes at a height where chickens cannot kick dirt and bedding into them. You can also cover feeders and waterers with screens to prevent this or—better yet —do both.

There are worm-preventive medicines and, if you know some of your birds have parasites, there are dewormers you can pick up at your local farm supply store. Be sure to check for egg withdrawal during the treatment window.

Fleas

Like most animals, chickens can fall prey to fleas. The type of flea that goes after them will not be the same as the ones that go for, say, the family dog, so you do not need to worry about cross-contamination between barnyard or backyard animals.

The good news is that fleas are relatively rare compared to lice and mites.

You'll know your chickens have fleas first and foremost because you will see the fleas crawling around on the birds. Other signs include bleeding, anemia, decreased egg production, weight loss, and missing feathers, but it takes a fairly severe infestation to reach the point of anemia.

There is a type of chicken-hungry fleas called *sticktight* fleas that attach to the chickens' skin and do not let go. Because they reproduce so quickly, this can turn into a serious infestation in a matter of a day or two.

Unfortunately, sticktight fleas *thrive* in settings like a

nice dust bath, which makes prevention especially difficult. These fleas do best in warm tropical climates so if you live in cooler areas you may not need to worry about them.

Whether the fleas are sticktight or regular, remember that the best prevention against fleas is a clean, dry coop. Checking your chickens daily for signs of trouble, and of course, providing routine dust baths will protect against all but sticktight fleas. For those, you can spray your chickens with a watered-down solution of apple cider vinegar to help prevent future infestations.

If your flock has fleas, you will need to get all the chickens out of the coop immediately and clean it thoroughly. Change out the bedding and spray everything down before you add fresh clean bedding for your chickens. Then dust the coop down in diatomaceous earth.

With the chickens themselves, remove any fleas (expect to need tweezers if you have sticktight fleas) and spray their bodies with a mixture of apple cider vinegar, taking care to avoid their faces. This should prevent further infestation, although if you've missed any fleas you may need to re-treat.

Injury

These can be caused by predators, household pets, or

even other chickens. They can range in severity from needing to be cleaned and dried to needing veterinary intervention to save the chick.

The most obvious preventative measures for this are to keep other animals away from your flock. Keep a lock on your coop, install motion-detector security cameras around your coop for 24/7 coop monitoring. You can add flashing around your yard to deter hawks and other predatory birds.

Watch for signs of aggressive behavior along the pecking order, and keep your chickens otherwise healthy so that no insect bite wounds get infected. Isolate any sick or injured birds to prevent worse pecking order injuries.

When your bird has an injury, you should first isolate the chicken and then wash and irrigate the wound. Assess it for damage. Some wounds, like a pulled feather, can be spectacular and messy but need relatively little care. Other wounds may seem small but are at high risk for infection due to being punctured.

Once the wound is clean, spray with 2% chlorhexidine solution, vetericyn, or betadine solution. Keep the chicken hydrated—this is more important than getting it food during the healing process, although you should offer food as well).

Keep the bird fully isolated until the wound has healed and then reintroduce to the flock carefully. Nighttime is the best reintroduction period: Add the bird to the flock before you close up the coop for the night. By morning, they won't even notice there is an extra among their numbers.

Marek's Disease

This is a highly contagious herpes virus that is widespread among poultry and other fowl throughout the world. It presents in different ways and can lead to immediate death or for a prolonged period of slow decline that takes years before the chicken dies.

An important note about Marek's Disease is that it can be vaccinated against and prevented. This, even if you are against other vaccines, is one that may be worth getting. Vaccinated chicks can even still be sent through the mail. It *must* be vaccinated against on the first day of life, however.

Marek's Disease typically strikes chickens between 12 and 30 weeks of age. It comes in four forms that impact different parts of the body: skin, neural, organs, and eye.

The dermatological, or skin, form causes enlarged feather follicles which can lead to scabbing and infection. This kind takes longer to kill the chicken.

The neural form, also called range paralysis, enlarges nerves. This results in paralysis in a leg, wing, or both. This kind kills fairly quickly, usually in a matter of a few days. It is the most common form of Marek's.

When Marek's Disease affects the organs, it causes diffuse tumors all over the body, including the liver, spleen, kidneys, heart, lungs, reproductive organs, stomach, and even in the muscle tissue. This takes a few days to a few weeks to kill and decline is usually rapid once tumors are apparent.

Because the tumors from the organ form of Marek's can occur in different parts of the body, they have a wide swath of symptoms, including kidney trouble, liver trouble, slowed crop movement, severe bloating, bleeding from the vent, and all sorts of other issues.

The final form of Marek's Disease, the ocular form, impacts the eye. It turns the eye gray and changes the shape.

While Marek's Disease is universally fatal, some birds can live for years with the milder forms (such as the ocular type). These birds will need to be isolated from your flock.

Upper Respiratory Infection

While this may call to mind the common cold, upper

respiratory infections in chickens can sometimes be caused by very dangerous germs. That is why it is so important to take preventive measures to protect your flock from this condition.

You'll want to keep coop bedding dry and be certain your coop provides adequate ventilation for all birds. Change the water daily, isolate new birds for 30 days before introducing them to your flock, and vaccinate whenever possible.

To avoid upper respiratory infections you absolutely *must* keep wild waterfowl away from your flock. This includes swans, geese, and ducks, all of which can carry some virulent germs to your birds.

Signs and symptoms of upper respiratory infection in chicken include sneezing, coughing, decreased activity, decreased egg production, and labored breathing. This is not a condition you will miss if you do a cursory inspection of your flock each morning: The warnings are quite evident from early in the infection.

Treatment depends on the cause of the infection, and unfortunately, this is often difficult to determine. You'll need the expertise of a veterinarian to help determine the cause.

In the meantime, isolate the bird in the sickbay area of your yard and check the remainder of your flock for signs of infection. Isolate any others that may have it.

At this point, all you can do (aside from condition-specific treatments prescribed by your vet) is to provide comfort care with electrolytes, water, and access to food. Your chickens need time and gentle coaxing to recover, though recovery is not a guarantee.

Sour Crop

This is a yeast infection in the crop, which is a digestive organ located in the chicken's chest. In a properly-working crop, food and scratch churn together to aid in the digestive process before the food moves into the stomach.

When the crop fails to drain, or when moldy food is introduced to the crop, yeast can start to grow. Not only will the colony and its waste fill-up the crop, but it can cause an 'overflow' from the crop back up the trachea and out the chicken's mouth. Any time you have a chicken with fluid draining from its beak, check the crop.

To avoid sour crop, you should not give your chickens starchy foods or anything that is clearly starting to rot or mold. You should keep their water clean—you can add apple cider vinegar to it to help maintain digestive pH balance—and you can add digestive aids to their food. The best for preventing sour crop include oregano, parsley, fennel, and garlic. Garlic should be given in moderation.

Another tip for preventing sour crop prevention is to keep the grass short in the areas where your chickens eat. Longer grasses can cause the crop to become impacted.

Signs and symptoms of sour crop include a swollen crop (it should be flat especially in the morning before they are given feed—if the crop is inflamed it will be easily-palpable through the chest wall), weight loss, gurgling sound in the chest, and fluid coming from the beak. Chickens are not able to vomit, so if you see fluid coming from the beak this is a flood of overflow fluids from a stalled crop.

To treat sour crop, massage the crop from top to bottom every two hours to try and work the food further down into the digestive tract. You should isolate the bird during the treatment process, and in this case, you want to prevent them from eating or drinking. After 12 hours (6 treatments) you can allow the chicken to have water, but not food.

If the crop becomes flat, that means it has emptied and you can now offer the chicken soft foods such as scrambled eggs, to help keep their system moving.

If the crop does not become flat, there is a procedure you can attempt. You should only do this when it is clear that the chicken is not recovering on its own, because it can trigger aspiration of fluid into the lungs, which can cause fatal pneumonia.

To expel the liquid *upward* from the crop:

1. Wrap the bird in a towel
2. Tip it upside down
3. Massage the crop from bottom to top for 15 to 20 seconds
4. If the chicken expels the contents of his or her crop, likely they will survive. Otherwise, they may not, but you should continue crop massages every two hours, as possible, to encourage a positive outcome.

Bumblefoot

This bacterial infection of your chicken's foot is more common in foot-feathered birds than in clean-legged birds, but can happen to any bird in any flock.

To prevent it, you'll want to check your flock's feet routinely for any early signs of infection. You should maintain a healthy nutrient-rich diet for your flock.

In the coop, you should make sure all roost bars are no

more than 18 inches from the floor of the coop, to prevent fall injuries. You should replace or repair broken roost bars or bars that have begun to shed splinters.

If your chicken develops bumblefoot, the most telltale sign is the large pus-filled abscess you'll find on the bottom of their foot when you investigate why they're limping or not moving at all. Their foot will be red and swollen, with a visibly-infected wound.

You may not enjoy reading about treatment for bumblefoot. If that is the case, skip over the remainder of this section and know that it is here if you ever need to know the treatment process.

Expect the treatment process to take an hour or more. First, you'll need to soak the foot in Epsom salt. When the flesh is more tender, use a very sharp knife to lance the abscess. Take care to gently separate the infected tissue from healthy, live skin.

From there on, you'll want to treat the foot like any other injury, and include wraps and antibiotic ointment as needed. Keep the chicken isolated from the rest of the flock until the foot fully heals, to prevent pecking order injuries.

Frostbite

Like any living organism, chickens are susceptible to the cold and the wet. Frostbite is a painful, unpleasant, and perfectly avoidable condition if you take proper care with your flock.

In colder months, you should ensure the chickens can keep their feet, comb, and wattle dry so that they cannot freeze.

Make sure that roost bars are flat rather than curved so that your flock can tuck their feet under the secure warmth of their feathers.

You can apply petroleum jelly to wattles, combs, and feet on the coldest days and nights. This will provide a protective layer not only to keep the skin dry but to prevent freezing.

If you live somewhere where the temperature frequently drops below 0 degrees Fahrenheit for days at a time in the winter, you should either insulate or heat your coop.

Take extra care to protect the feet of your flock, as they are the most susceptible to risk for frostbite.

Signs of frostbite in your flock include loss of part or all of the comb, wattle, or toes, and discoloration to those pieces of skin in less severe cases. Immediate treatment is necessary: Immerse or rub the frostbitten area with water around 100 degrees Fahrenheit for 15 to 20 minutes. Coat the area in petroleum jelly to prevent further damage.

Prolapsed Vent

As your hens get older, they may develop a prolapse in their vent from years of egg-laying. This can be treated surgically, but in some cases, birds will need to be culled if the prolapse is severe enough. There is no prevention, and no treatment outside of surgical intervention or end of life planning.

8

TABLE FARE

Whether you have layers for eggs, broilers for meat, or dual-purpose birds you intend to use to feed your family for the coming years, there is a lot to know as you embark on this amazing journey of sustainability and cruelty-free chicken farming.

AN EGGSELLENT MEAL

Backyard-grown fresh farm eggs are one of the most delicious things you can ever grow for yourself. They tend to have a stronger (but pleasant) taste than store-bought eggs due to the chickens having a richer diet, and they come with the added satisfaction that you know these eggs:

You know where they came from, what their layers ate.

You know how old they are, how long they've sat in your fridge. You know how happy the hens that laid them are, and most of all, you know you are giving a good home to chickens that might not have that opportunity without you.

When you sit down to a meal cooked using your homegrown eggs, you deserve all the good feelings and enjoyment that come with knowing you've done this all yourself.

There are things you can do, and information you should know, to make this experience as good as possible for you and your family (Martinko, 2018).

The first tip is to crack open each egg separately in its own preparation dish before adding it to the main dish. This is good practice anyway, but on the off chance you have a layer with a parasite you do not want to discover that *after* the worm has been added to your baking project.

Eggs removed from the coop should not be washed immediately, as it reduces shelf life. They can be stored outside the kitchen, to reduce the risk of germ spread, and washed when you are ready to use or sell them.

Some people like to keep a separate minifridge for their farm fresh eggs, in the interest of improving shelf life. Since unwashed eggs last longest, and since a day on the counter is equivalent to about a week in the fridge, having a dedicated minifridge for your eggs will reduce the spread of bacteria and also extend your eggs' shelf life.

Always be sure to cook eggs thoroughly to reduce the risk of salmonella contamination. Similarly, it is important to wipe down the kitchen area with a disinfectant spray after cooking with raw eggs. Doing so will help to keep you and your family healthy and happy.

EGG SALES

S. Hermann & F. Richter Four Eggs (2019) retrieved from www.pixabay.com

Regulations for egg sales vary from state to state, and even from county to county within a state. Most areas allow backyard chicken farmers to sell farm fresh eggs. There may be regulations concerning labeling, pricing, and cleanliness of the eggs.

Remember that eggs have a longer shelf life if the protective layer is not washed off them. Some areas will require you to wash the eggs before you can sell them.

You must do your research to protect yourself and your family from any fines due to misunderstandings that crop up when it comes to egg sales.

MEAT PROCESSING AND COOKING

Eating a home-raised chicken can be a rewarding experience, especially if the meat is processed right to ensure tenderness and preservation of flavor (McCrea, 2017).

Before you cull, you will want to stop access to feed at least four hours ahead of time. This will help empty the intestines and prevent the chance of mistakenly spilling the contents of the intestines on the meat.

The most effective way to cull a chicken is through removing its head. To keep it as humane as possible, you will want to make sure your blade is as sharp as you can manage. Another option, which will reduce bruising to the meat, is to hang the bird upside down and snick the carotid artery with a sharp knife.

If you remove the head, you will need to hang the chicken after that step to drain the blood. The better-drained the chicken meat is, the better the flavor will be. Blood can leave behind a metallic taste that even the best herbs and spices cannot cover-up.

While the blood is draining, you can push a knife through the roof of the mouth and into the brain. This process, called *pithing,* is designed to hit the hypothalamus and cause the chicken's body to reflexively let go of all the feathers. You want to angle the knife through a slit in the roof of the mouth and toward the eye. It is not guaranteed to trigger the release of feathers, but it is certainly helpful when it works!

Once the chicken has drained, take a large pan that is big enough to fit the whole chicken. You will want to fill it with enough water to submerge the carcass. Heat the water (not to a boil) and scald the meat. This will make the feathers easier to pluck.

You can tell a chicken is done being scalded by picking at toe scales and toenails. If they slough off easily, the chicken is ready to be removed from the water. Be sure to remove all toenails and scales before you start on the plumage.

When removing the feathers, hold the chicken carcass steadily and pull in a downward motion relative to the slant of the feathers. This process can be time-consuming, and if you plan to do long-term meat culling in your life you may benefit from purchasing an automatic plucker to do this part of the work for you.

To remove the organs, start by cutting the flesh around the vent in a circle to release the vent from the flesh of the chicken. Once this is done, use your fingertips to widen the opening around the intestines.

From the other end of the bird, use clippers to remove the head if it was not removed during the cull. With a sharp knife, slice along the larynx to create an opening. Use your fingers to gently pry the esophagus loose from the flesh around it. From there, work your way down to the crop and work it loose from the surrounding tissue. The crop will resist your efforts, so be prepared for a bit of a struggle there.

From the vent end, reach inside the chicken and pull out the heart and the remaining internal organs, taking care not to crush or 'open' any inside the chicken. If you do, the bird will need to be rinsed thoroughly before you continue.

At the base of the tails is a gland called the *preen* gland, which will need to be cut out of the chicken so that it does not alter the flavor.

Lastly, remove the feet at the place where scales stop and bare flesh begins.

Rinse the chicken vigorously and then place it in an ice-water bath large enough to allow for full immersion. Let it sit for four hours, to help the chicken's body process rigor mortis and to keep the meat as tender as possible. Use a thermometer to ensure that the thickest part of the chicken—usually the breast—reaches or drops below 40 degrees Fahrenheit.

Drain the carcass again and place it in a resealable plastic bag, to be refrigerated or frozen until you are ready to use the meat.

Enjoy your meal, when the time comes! You've earned it!

COOKING TIPS

Lebensmittelfotos Chicken Carcass (2019) retrieved from www.pixabay.com

Broiler birds are so naturally flavorful that many people subscribe to the school of thought that the meat should speak for itself when you cook it. Even dual-purpose laying hens, if prepared right, can yield a tender and delicious meat that needs little to no flavoring.

You have a range of options for how to cook the meat to your liking. Some people invest in a rotisserie and cook the entire chicken in one go. Others choose to boil the carcass in a mixture of select herbs and spices to create their own homemade chicken stock. They separate the bones and use the chicken in other dishes, soup, or sandwiches in the coming week.

In the spirit of sustainability, you will want to waste as little of the chicken as possible. Bones can be softened in a pressure cooker after they are used for stock, and these softened bones can be crushed and fed to dogs or cats for protein intake.

CONCLUSION

From the earliest musings that you might get a handful of chickens to make your backyard more sustainable, to the day you realize you're an experienced backyard farmer with the confidence to care for your flock, there is no doubt that backyard chicken farming is a journey.

It is a journey of information, knowledge, experience, bonding, and growing. There is no doubt that caring for these creatures will change how you see the world and change your role *in* the world.

A quick summary of the most critical things to remember while you're preparing for this journey:

- Coops do not need to be beautiful or expensive to serve the needs of your flock.

- Clean food and water, and clean coops, lead to healthy chickens.
- Whatever unexpected twists you face in the road, know that you are capable of addressing them and serving your flock well.
- Having a sickbay and chicken first aid kit on hand will prepare you for whatever you encounter on this adventure.

Harald Dona, Painting of Chickens (2019) retrieved from www.pixabay.com

The biggest takeaway from backyard farming is to follow your instincts and principles, to have fun, and—

perhaps most important of all—to eat well. Happy farming!

DISCUSSION

CHICKENS BY THE NUMBERS

For a direct comparison of each breed of chicken and what it does best, please use this chart. While it doesn't address temperament or hardiness, it gives a good idea of what sort of outcome you can expect from each bird.

	Eggs			Size (lbs)		Type	
	Annual	Weekly	Color	Roo	Hen	Layer	Broiler
Australorp	250	5	brown	8	6.5	X	––
Barnevelders	200	3-4	lt brown	6	5	X	––
Brahmas	200	3-4	brown	10	8	X	X
Bresse	––	––	––	6.5	6.5	––	X
Cornish Cross	––	––	––	11	9	––	X
Freedom Ranger	––	––	––	6	6	––	X
Isa Brown	350	6-7	red-brn	5	5	X	––
Jersey Giant	––	––	––	12	6.5	––	X
Leghorn	300	5-6	white	7.5	6	X	––
Marans	150	3	brown	8	6.5	X	––
Orpington	––	––	––	8	7	––	X
Plymouth Rock	200	4-5	brown	8	7	X	––
Rhode Island Reds	250–300	5-6	brown	8.5	6.5	X	––
Sussex	250	5	lt brown	9	7	X	––
Wyandottes	200	4	lt brown	9	7	X	––

GLOSSARY OF TERMS

Apple cider vinegar — a type of acid which is useful for treating and preventing a variety of ailments

Automatic Door — opens and closes on its own, frequently scheduled for dusk and dawn to keep chickens safe

Bantam — a smaller version of chicken breeds, typically ½ to ⅔ the size of a full-grown chicken, lays smaller eggs

Betadine — a disinfectant safe to use with chickens

Botflies — flies that lay their eggs in the flesh of their prey so that their maggots have something to eat

Broiler — a bird grown and raised specifically for table fare

Brooder — a safe, heated area where baby chicks can grow and live until they are ready to move to the coop

Broody Hen — a hen that will not leave her nesting box because she has decided to grow a fresh batch of chicks

Bumblefoot — an infection in the foot or toes of a chicken

Caseous Exudate — an egg-shaped deposit full of bits of egg, fatty tissue, and other gunk

Chicken Math — the idea that a flock will grow in unexpected ways

Coccidiosis — an intestinal parasite that affects chickens and can be fatal

Cockerel — a young male chicken

Cockfighting — an illegal practice you may want to be wary of when rehoming roosters

Comb — the crest, a growth of flesh on the center top of a chicken's forehead

Compositing — the practice of turning food and garden waste into fresh soil

Coop — the physical housing structure chickens live in

Crop — a small area in the breast of the chicken where food goes shortly after it is eaten

Crushed Oyster — crumbles of calcium-heavy scratch for your chickens

Cull — for behavioral, medical, or feeding purposes, remove some birds from your flock by terminating their life

Diatomaceous Earth — a powdered sedimentary rock useful in dust baths for preventing a variety of chicken ailments

Dual-Purpose — a bird that can be used as a layer and retired as a broiler

Dust Bath — dry sand with ashes or diatomaceous earth which chickens can clean themselves in and which helps prevent external parasites

Egg Withdrawal — some medicines chickens need explicitly state that the chickens' eggs should not be consumed following treatment — sometimes temporarily and other times for the remainder of the hen's life

Electrolytes — minerals and salts that help living organisms function

Factory Farm — a farm designed to process broiler birds and layers without attention to the emotional wellbeing of the birds

Feather Sexing — a method of determining the gender of very young chicks based on the wing growth pattern

Feeder — a container for providing food to your flock

Fleas — an external parasite that can infest your chickens

Flystrike — this is the word for a botfly attack, which can be fatal

Free-Range — allowing your chickens to roam your yard with few to no restrictions

Frostbite — tissue damage that occurs when flesh becomes wet and then freezes

Hardware Cloth — mesh screening material that helps deter predators around a coop

Hatchery — a place where baby chicks are born and sold to farmers

Heat Lamp — a red-tinted lamp designed to heat a

brooder box without disrupting the chicks' circadian schedule

Hen — a mature female chicken

Incubator — an electronically heated device designed for hatching fertilized eggs

Lash Egg — soft egg caused by not enough calcium in the chick's diet

Layer — a hen kept specifically for her egg-laying abilities

Layer Feed — food fortified with nutrients specifically beneficial to laying hens

Lethargy — severe decrease in activity, usually with sudden onset, indicates a chicken is ill

Lice — an external parasite chickens are prone to, not the same as human head lice

Lockdown — the final three days of the incubation period, during which eggs should not be disturbed

Marek's Disease — a neurological disease which can be prevented by vaccination

Mealworms — a special treat that chickens love and you can grow yourself if desired

Medicated Feed — chick feed containing a medicine to help build resistance against coccidiosis

Mites — an external parasite that can infest your flock

Molt — around eighteen months of age, chickens lose their feathers and grow new ones

Pecking Order — the established social hierarchy of a flock, can change frequently and result in fights between birds

Pip — break out of the egg in order to hatch

Pith — insert a knife into the bird's brain in an effort to get the hypothalamus to release all the feathers

Pluck — remove the feathers from a carcass

Prolapsed vent — a failure of the tissues around a vent

Pullet — a young female chicken

Quarantine — isolate one or more members of a flock to prevent or limit the spread of disease

Raised Coop — sits off the ground to prevent rot and infestation

Rehome — find a better living situation for some or all of your birds, new chicks, or spare roosters you have

Roost Bar — a place for chickens to sleep off the ground

Rooster — a mature male chicken

Run — a safe enclosed space where chickens can move around freely and forage

Scald — heat the surface of something rapidly, without boiling

Scraps — leftovers from your own dinner table and food preparation process, which can be fed to the flock

Scratch — gravel and rocks used to help chickens move food through their crops

Security System — a remote access camera that can be used to keep an eye on your flock and determine what predators, if any, are targeting your ladies

Semi-permanent Coop — chicken housing you move around your yard occasionally or seasonally to benefit the flock

Sickbay — a place to isolate sick or injured chickens during the course of medical intervention

Socialization — is important between you and all your birds, to establish rapport and to be able to handle your chickens without complaint when it is medically necessary

Soft Shell – in an egg, indicates a lack of calcium

Sour Crop – a yeast infection in the crop which is caused by delayed crop emptying

Starter Feed – food specially designed for newborn and growing chicks

Stationary Coop – chicken housing that cannot be moved

Straight Run – ungendered chicks

Supplemental Daylight – may be beneficial to year-round layers in winter

Thin Shell – a finding in some eggs, and an indication that your birds are short on some key nutrients, most notably calcium

Tractor Coop – a chicken housing structure that can be moved around the yard as needed

Upper Respiratory Infection – a cold, bronchitis, or influenza that can infect chickens

Vaccination – the practice of exposing healthy birds to an attenuated version of an illness to encourage immunity in the bird

Vent – the opening through which refuse and eggs pass out of the chicken

Vent Sexing — a method of determining chick gender by examining the vent

Vetericyn — a helpful spray and ointment worth having on hand for the days when you need to reread this guide and discover what to do to help your sick or injured bird

Waterer — a container designed to provide water to chickens and their young

Wattle — flaps of flesh at the chicken's neck

Worms — an internal parasite that, if left untreated, can lead to death

Wry-Neck — a condition of numerous possible causes, in which the chicken's neck is unable to support its head

CHICKEN FIRST AID KIT

It is important to have a ready-made, easy-access first aid kit stored in or near your coop (for temperature regulation, it can be somewhere in the house if necessary, so long as it is easy to find when you need it.

Your first aid kit should contain all of the following:

- petroleum jelly

- vetericyn spray and ointment
- poly vi sol
- tweezers
- medical tape
- probiotics
- apple cider vinegar
- electrolytes
- nail trimmers
- scissors
- cornstarch
- nitrile gloves
- diatomaceous earth
- 2% chlorhexidine solution

The Beginner Gardener's Tool Checklist

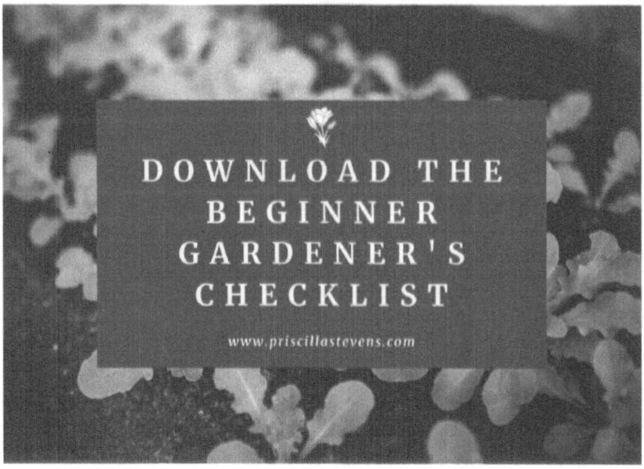

A *must* read before you start your garden!

This Checklist Includes:

- 7 items you can't start a garden without.
- The highest quality items.
- Where you can buy those items for the lowest price.

Don't get overwhelmed when starting your first garden!

To receive your checklist, scan the QR code below!

REFERENCES

A, T. (2013, December 28). Must-Have Supplies To Get Before You Bring Your Chickens Home. Retrieved May 9, 2020, from https://www.offthegridnews.com/how-to-2/must-have-supplies-to-get-before-you-bring-your-chickens-home/

B., E. (2019, April 17). Myth-Busting Medicated Chicken Feed: Feeds That Include Amprolium. Retrieved May 11, 2020, from http://www.scoopfromthecoop.com/myth-busting-medicated-chicken-feed/

Backyard Chickens. (2020). Hatching Eggs & Raising Chickens. Retrieved May 9, 2020, from https://www.backyardchickens.com/articles/categories/hatching-eggs-raising-chickens.22/

Caldwell, N. (2013, March 14). Benefits of Backyard Eggs. Retrieved May 1, 2020, from https://www.motherearthnews.com/homesteading-and-livestock/benefits-of-backyard-eggs/

Cackle Hatchery. (2018, July 3). Should You Vaccinate Your Chickens Against Marek's Disease?: Cackle Hatchery. Retrieved May 11, 2020, from https://blog.cacklehatchery.com/should-you-vaccinate-your-chickens-against-mareks-disease/

Chicken Keeper. (2013, May 13). Chicken Care: Daily Routines. Retrieved May 13, 2020, from http://achickenkeepersblog.blogspot.com/2013/05/daily-routines.html

Elise. (2019, April 17). Free Ranging Chickens: The Pros and Cons. Retrieved May 11, 2020, from http://www.scoopfromthecoop.com/pros-and-cons-of-free-ranging-chickens/

Ersek, K. (2012, September 11). The 6 Essential Nutrients for Healthy Plants. Retrieved May 2, 2020, from https://www.holganix.com/blog/the-6-essential-nutrients-for-healthy-plants

Green Willow Homestead. (2017, December 14). The Pros and Cons of Having a Rooster. Retrieved May 11, 2020, from https://www.greenwillowhomestead.com/blog/the-pros-and-cons-of-having-a-rooster

Griffler, Z., & Griffler, Z. (2020, February 14). How Chickens Get Along With Other Species. Retrieved May 12, 2020, from https://opensanctuary.org/article/chickens-and-other-species/

Happy Chicken Coop. (2017, July 24). The Happy Chicken Coop. Retrieved May 14, 2020, from https://www.thehappychickencoop.com/buff-orpington/

Happy Chicken Coop. (2017, August 18). The Happy Chicken Coop. Retrieved May 14, 2020, from https://www.thehappychickencoop.com/jersey-giant/

Happy Chicken Coop. (2017, November 14). The Happy Chicken Coop. Retrieved May 6, 2020, from https://www.thehappychickencoop.com/rhode-island-red/

Happy Chicken Coop. (2017, December 14). The Happy Chicken Coop. Retrieved May 3, 2020, from https://www.thehappychickencoop.com/australorp-chickens-a-comprehensive-care-guide/

Happy Chicken Coop. (2018, January 31). The Happy Chicken Coop. Retrieved May 6, 2020, from https://www.thehappychickencoop.com/black-copper-marans/

Happy Chicken Coop. (2018, March 31). The Happy Chicken Coop. Retrieved May 6, 2020, from https://www.thehappychickencoop.com/sussex-chicken/

Happy Chicken Coop. (2018, April 23). The Happy Chicken Coop. Retrieved May 5, 2020, from https://www.thehappychickencoop.com/ISA-brown/

Happy Chicken Coop. (2018, May 10). The Happy Chicken Coop. Retrieved May 6, 2020, from https://www.thehappychickencoop.com/leghorn-chicken/

Happy Chicken Coop. (2018, June 25). The Happy Chicken Coop. Retrieved from https://www.thehappychickencoop.com/wyandotte-chicken/

Happy Chicken Coop. (2018, July 19). The Happy Chicken Coop. Retrieved May 3, 2020, from https://www.thehappychickencoop.com/barnevelder-chicken/

Happy Chicken Coop. (2018, October 30). The Happy Chicken Coop. Retrieved May 6, 2020, from https://www.thehappychickencoop.com/plymouth-rock-chicken/

Happy Chicken Coop. (2019, January 2). The Happy Chicken Coop. Retrieved May 14, 2020, from https://www.thehappychickencoop.com/red-ranger-chicken/

Happy Chicken Coop. (2019, June 27). The Happy Chicken Coop. Retrieved May 14, 2020, from https://www.thehappychickencoop.com/bresse-chicken/

Happy Chicken Coop. (2019, July 5). The Happy Chicken Coop. Retrieved May 14, 2020, from https://www.thehappychickencoop.com/cornish-chicken/

Happy Chicken Coop. (2019, December 13). The Happy Chicken Coop. Retrieved May 1, 2020, from https://www.thehappychickencoop.com/keeping-backyard-chickens-what-i-wish-id-known/

Harvill Homesteader, J., & Gardener, L. N. L. (2019, January 29). Turn Chicken Poop Into Garden Gold. Retrieved May 2, 2020, from https://www.epicgardening.com/chicken-manure/

Hassan, S. M., Dipeolu, O. O., Amoo, A. O., & Odhiambo, T. R. (1991, March). Predation on livestock ticks by chickens. Retrieved May 1, 2020, from https://www.ncbi.nlm.nih.gov/pubmed/1858289

Kathy. (2014, June 18). 10 Tips for Healthy Chickens. Retrieved May 10, 2020, from https://www.communitychickens.com/10-tips-for-healthy-chickens/

Martinko, K. (2018, October 11). Are backyard eggs really that dangerous? Retrieved May 14, 2020, from https://www.treehugger.com/green-food/backyard-eggs-carry-risk-salmonella.html

McCrea, B. (2017, September 12). How to Process

Chickens At Home. Retrieved May 11, 2020, from https://www.chickenwhisperermagazine.com/health-and-wellness/how-to-process-chickens-at-home

Mormino, K. S. (2017, November 27). Egg Oddities: A Guide To Irregular Eggs. Retrieved May 8, 2020, from https://www.hobbyfarms.com/irregular-eggs-unusual-chickens/

Price, A. (2019, July 18). Conventional Chicken Messes with Your Hormones & Could Cause PCOS, a Leading Cause of Infertility in the U.S. Retrieved May 1, 2020, from https://draxe.com/nutrition/free-range-chicken/

Ramsey, L. (2018, February 23). These Are The Top 15 Deadliest Animals on Earth. Retrieved May 1, 2020, from https://www.sciencealert.com/what-are-the-worlds-15-deadliest-animals

Stratton, A. E. (2013, June 26). 10 Ways to Build a Better Chicken Coop. Retrieved May 10, 2020, from https://www.thisoldhouse.com/pets/21018574/10-ways-to-build-a-better-chicken-coop

Timber Creek Farm. (2016, August 19). What Kind of Chicken Coop Design is Best? Retrieved May 10, 2020, from https://timbercreekfarmer.com/what-kind-of-chicken-coop-design-is-best/

West, M. (2017, July 17). The Benefits of Keeping

Chickens. Retrieved May 3, 2020, from https://www.permaculturenews.org/2017/07/18/benefits-keeping-chickens/

Weston, L., Burdett, A., Beck, D., & Wolff, R. (2018, November 3). Backyard Chicken Health Issues: Symptoms Treatment & Prevention. Retrieved May 11, 2020, from https://www.caringpets.org/how-to-take-care-of-a-backyard-chicken-hen/health-issues/

www.ingramcontent.com/pod-product-compliance
Lightning Source LLC
LaVergne TN
LVHW091531200926
848483LV00002B/360

* 9 7 8 1 7 7 7 3 9 8 1 2 5 *